HOUSE

BUILDING & REMODELING IDEAS

YOUR ENERGY EFFICIENT HOUSE

BUILDING & REMODELING IDEAS

Anthony Adams

GARDEN WAY PUBLISHING
Charlotte, Vt.

Illustrations by Anthony Adams

Designed by Henry Huston

Printed in the United States by Capital City Press and published by Garden Way Publishing, Charlotte, Vermont 05445.

Third Printing, June 1977

Library of Congress Cataloging in Publication Data

Adams, Anthony, 1936-
  Your energy-efficient house.

  Bibliography: p.
  1. House construction. 2. Dwellings—Remodeling.
3. Energy conservation. I. Title.
TH4811.A3   690.8   75-16829
ISBN 0-88266-073-X
ISBN 0-88266-053-5 pbk.

Contents

v

PROLOGUE

There was a time when "energy" referred mainly to the physical efforts of an individual, his family, servants and animals. It meant, too, the heat and light generated by burning the products of nearby fields and forests. It was a simple energy "system," close to nature, requiring virtually no cash to maintain it, although it took much physical effort.

Shelter then was planned carefully with a view to climate, the characteristics of location, specific needs of those to be sheltered and the materials available to construct it. The sources of "energy" available nearby were adequate to support such shelter, mostly because the two were considered together.

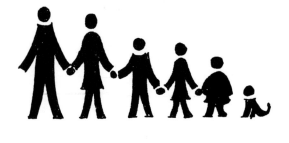

Today energy is something derived from afar, to be purchased at great cost and transported at great expense to be ultimately used in great amounts.

Such great amounts are required because activities are conceived without regard for the energy's uses and its cost, particularly in the building of shelter. Buildings are built without regard to the climate, in locations not particularly suitable to habitation, and in ways that have little to do with the needs of those to be sheltered.

And now the dream of a way of life based on abundant, cheap, remote energy has become a nightmare in terms of availability and personal expense.

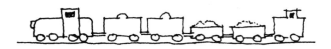

1

WHO ARE YOU?

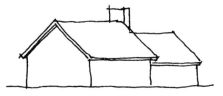

This book is about building, mostly about the places we live in or would like to live in, but it can also apply to other spaces that we enclose by structures. The book is also about energy and the ways we use it or, more importantly, don't use it, particularly in buildings.

Perhaps most important, the book is about you, your family needs and desires, about the place you live, and spend a great deal of your time, and perhaps also where you work.

The intent of the following words, pictures and diagrams is to acquaint you with the many opportunities that are available to provide comfortable and satisfying sites and buildings in which to live or spend your leisure or working time. The possibilities emphasize developing more self-sufficiency and a lesser need for unreliable and expensive outside sources of energy, of food and even of things to do.

The ultimate purpose, then, is to help reverse the present trend of using and spending more and still more, and to start spending less.

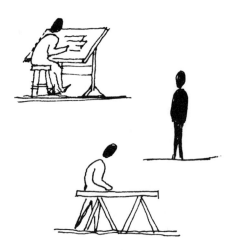

This book is not about super solutions, nor yet is it theoretical, nor wishful. It is about your own neighborhood, home and family, and about what you can do right now to live better and more economically.

If you can supply a modest amount of background knowledge, a desire to do something and a few ideas of what you want, we can help you to provide what you need. In making these suggestions we assume your thought and intuition are the cheapest commodities available, and we propose that you use them as central to the process. We will help develop your ideas and assist in putting them into effect.

What should we call this process?

If you are a business or governmental executive it is called planning, if you are an architect or engineer it's design. To a carpenter or pipefitter, it's craftsmanship (or good workmanship). Lacking the credentials to use any of the above terms, it becomes common sense.

A word about who this book is for:

It is for the home owner concerned about spending so much money on energy and perplexed as to what to do about it.

It is also for the person who has a home but wants to add or renovate. This may include minor alterations, a major addition or the total recycling of a decrepit or otherwise inadequate existing building.

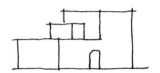

It is also for somebody who wants to make a beginning, whether it is the first house owned, a desire for a whole new lifestyle or the realization of a lifetime desire to build "your own place." Even for the person who wants to experiment in new ideas and new technology in housing, this will serve as a useful start.

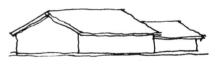

No matter how large or small in size, near or distant in time, or costly or inexpensive your proposed project, the important thing is the way you go about it.

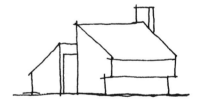

The following pages can provide you with a framework on which you can build toward your goals. Though all of the things discussed may not apply directly, most of them will relate in some way to things you will do, or will help you to evaluate circumstances that already exist.

Perhaps a diagram will help explain. We can list all the steps to be considered to create an energy-efficient house.

1

Next we can show where you fit into or enter the process, whether you are starting at the beginning, doing renovation or doing some periodic maintenance and yard work. The steps are:

1. Finding and evaluating a site for your new home.

2. Planning the use of a site — even one you already own or occupy.

3. Deciding the uses, character and layout of spaces in your new or remodelled home.

4. Determining the size and shape for a building, how it is to work for you, as well as how it will look to you and others.

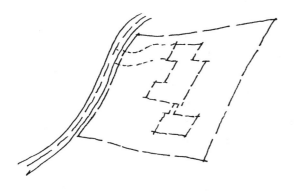

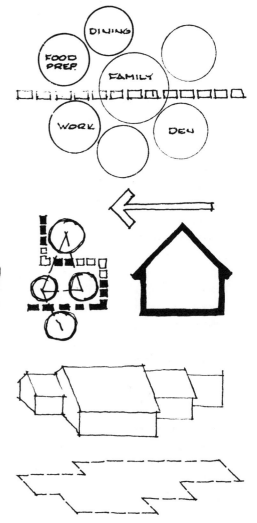

5. Figuring out room sizes and features, such as window and door openings.

6. Choosing structural systems to support the building, and materials to enclose spaces that will serve your needs and desires.

7. Selecting building systems — that is heating, lighting, water supply, waste disposal and other systems.

8. Designing exterior elements of your home to derive the greatest economy, the best use and personal enjoyment.

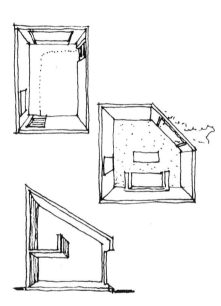

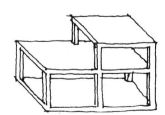

The way you fit into the program can be shown as follows:

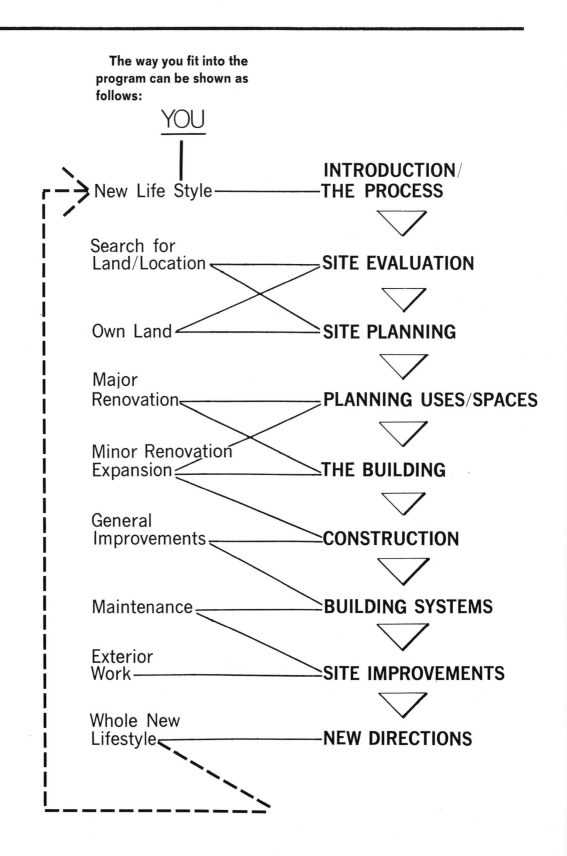

YOU

New Life Style ———— INTRODUCTION/THE PROCESS

Search for Land/Location ———— SITE EVALUATION

Own Land ———— SITE PLANNING

Major Renovation ———— PLANNING USES/SPACES

Minor Renovation Expansion ———— THE BUILDING

General Improvements ———— CONSTRUCTION

Maintenance ———— BUILDING SYSTEMS

Exterior Work ———— SITE IMPROVEMENTS

Whole New Lifestyle ———— NEW DIRECTIONS

NEW	INTRODUCTION/	RECYCLED

NEW		RECYCLED
A Beginning————————	**INTRODUCTION/ THE PROCESS**	
	▽	
Looking for Land/Location	**SITE EVALUATION**	Examine Surroundings
	▽	
Own Site	**SITE PLANNING**	New Potentials
	▽	
Your Needs/Desires	**PLANNING USES/SPACES**	Rethink Spaces/ Living Patterns
	▽	
Uses/Spaces, a New Building	**THE BUILDING**	New Uses/a New Presence
	▽	
Making Choices/ Establishing Directions	**CONSTRUCTION**	Compatibility/ Continuity
	▽	
Veins, Arteries, Nerves	**BUILDING SYSTEMS**	Replace/Update
	▽	
The Outside Spaces	**SITES IMPROVEMENTS**	More New Potentials
	▽	
Future Projects	**NEW DIRECTIONS**	Future Projects

SITE SELECTION & EVALUATION

The first consideration of your energy-efficient house is a place to build it — or the place where it already stands.

If you don't yet have a site, the following points will help you in selecting one. If you have a number of sites in mind, this can help you choose between them. If you already own land for building, the following will help you see it in terms of how you can use it, and the ways you will require energy to live there.

If you are already living on the site you can examine your existing home and site and evaluate how well you are using it, particularly in regard to the use of energy. Using these methods you will be able to see how well you are doing, or how you might do better. You may even decide that your present situation is hopeless to make better, although we would like not to think so.

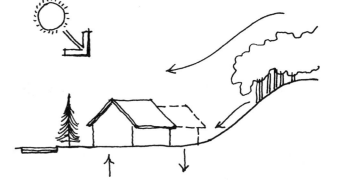

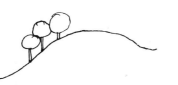

In saving energy and money you should be concerned first with climatic factors, with the sun and wind and their potential for providing energy. These are factors that are generally fixed for any region or geographic area.

Of major concern are the ways in which sun and wind affect a given site, the ways in which they may be modified by site characteristics and the ways in which you may adapt effectively to them.

Some of the site factors to evaluate are soils, slope and overall topography, existing vegetation and nearby bodies of water.

We will not be concerned here with the location as it relates to your work or to schools, nor with land costs. These factors are adequately dealt with in books cited in the Reading List.

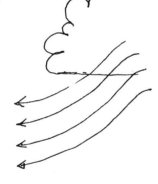

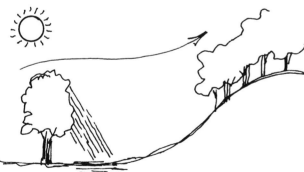

THE SUN

One of two major factors influencing the energy used on any site is the sun. The other is wind. The sun delivers wanted energy free of charge during cool periods, but it also at times provides excess heat which may be removed at great cost in energy.

But let's look at the simple things first.

While the sun always rises, reaches its zenith and sets in the same points of the compass, it varies geographically, being higher in the sky in southern than in northern latitudes. Season variation also occurs, of course, because of the precession of the earth's axis, and the sun is higher in the sky during the summer and lower in winter (reverse in the southern hemisphere). So much for the sun.

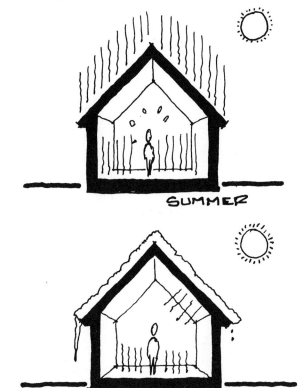

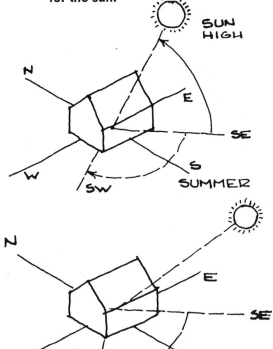

The second major factor, the wind, deprives you of heat during cool periods but is most helpful in removing unwanted heat in times of excess warmth.

In our hemisphere the prevailing winds generally blow from west to east. But seasonal and regional variations may find warm and often moist winds blowing from the south, and cold, dry winds from the north.

These patterns combine to establish southwesterly and northwesterly winds which often are identified with a certain season.

Other wind patterns derive from the weather characteristics of a region. Thus we sometimes experience cold blizzards from the north and northeasters from the ocean.

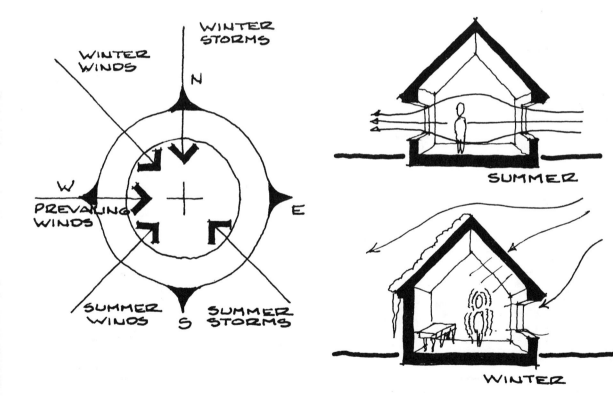

MORE WIND FACTORS

Wind patterns are affected also by large geographic features. Major mountain ranges will alter the direction of prevailing winds; large bodies of water will make them warmer or colder; large open expanses will allow them to reach full velocity; broken terrain will dissipate or scatter these effects.

It is important to know about all these variations of wind patterns where you plan to settle. The local Weather Bureau may help establish them, as will conversations with people whose livelihood depends on weather (farmers, fishermen, ranchers, woodsmen).

Using this wind information and a simple pocket compass you will be able to make some highly-educated guesses as to how overall climate condition will affect your potential home.

Your localized summary may look like this:

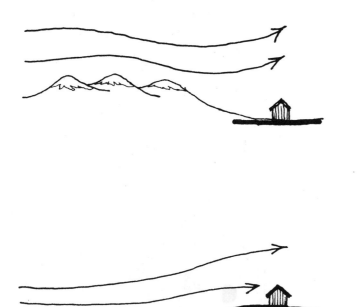

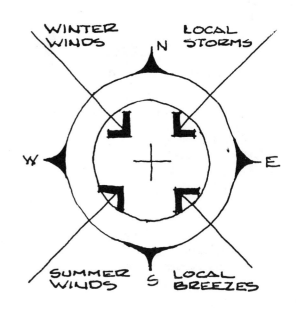

Your next step is to see how the physical features of a site will be affected by or perhaps will modify general climatic conditions.

A surface most nearly to a right angle to the sun's rays will absorb the most energy from the sun. Thus, given something less than a sun directly overhead, a sloped ground surface will absorb more than a flat one. When the sun is at its highest point, it will be most nearly at a right angle to the ground, so a slope facing south (in the Northern Hemisphere) absorbs most of all. Slopes that face east and west therefore will get less heat from the sun, and one facing north will get the least. Flat land makes little discernible difference. So much again for the sun.

Surfaces directly exposed to the wind will receive its full force. Surfaces sloping toward the wind will be all the more affected. Exposure is made more severe as wind velocities are increased in being compressed by the topography. Steeply-sloped or broken terrain will cause turbulence or eddies. A site on broad expanses of flat area, such as prairies or adjacent to lakes or oceans, are exposed to the full sweep of the winds.

MOST   LESS   LEAST

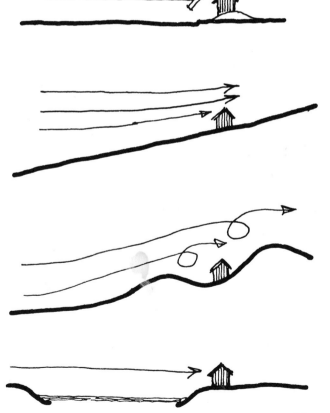

CONCLUSIONS – CLIMATE

Sun

South-facing building sites are warm.
North-facing building sites are cold.
East-facing building sites get early morning, warming sun.
West-facing building sites often get hot, afternoon sun.

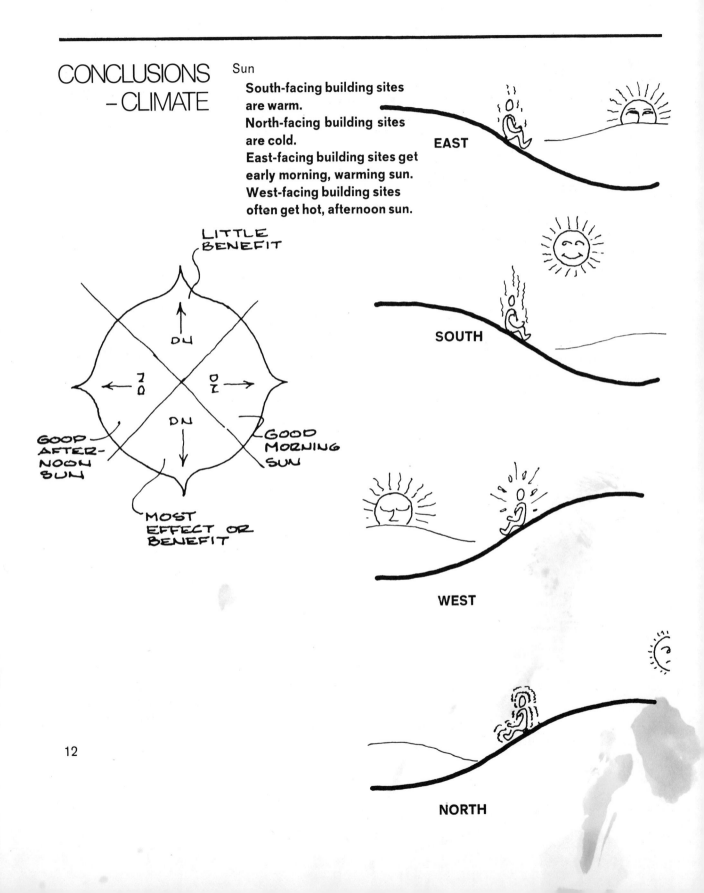

LITTLE BENEFIT

GOOD AFTERNOON SUN

GOOD MORNING SUN

MOST EFFECT OR BENEFIT

EAST

SOUTH

WEST

NORTH

Wind

Hill tops, ridges, profiles and higher elevations have more exposure to the wind, and will be colder in cold seasons, cooler in warm.

Slopes facing away from prevailing winds are sheltered.

Slopes facing toward prevailing winds are cooler. Valley bottoms will be warmer in the warmer seasons, cooler in cold seasons.

Sites on flat expanses are open to full sweeps of winds. Air temperature near large bodies of water will be tempered by the wind.

On a smaller scale a mountain, hills or even smaller knolls will provide shelter from a cold wind on the down-wind side, or will effectively cut off the refreshing coolness of a summer breeze.

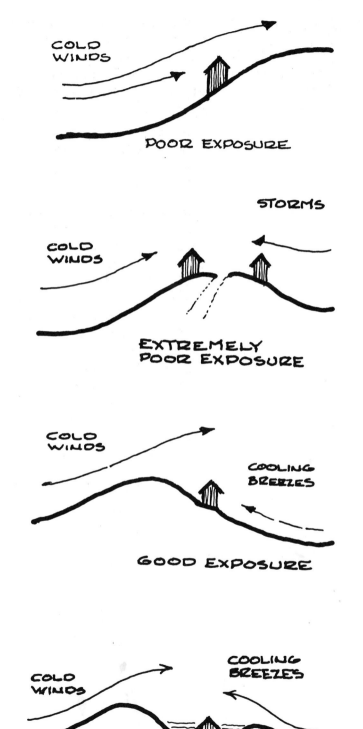

COLD WINDS

POOR EXPOSURE

STORMS

COLD WINDS

EXTREMELY POOR EXPOSURE

COLD WINDS

COOLING BREEZES

GOOD EXPOSURE

COLD WINDS

COOLING BREEZES

13

CONCLUSIONS – CLIMATE & VEGETATION

Vegetation can be considered as a shelter, too, or a way of modifying wind currents. Here again is a seasonal variation which can be harmful or helpful depending on how you use it.

Deciduous trees are good for summer shelter from the wind.

Evergreen trees are good for shelter in all seasons.

Deciduous trees, having lost their leaves in the fall, are not particularly good for winter shelter purposes. But a whole grove of them is fairly effective even in winter. They are good for summer wind protection, if desired, and excellent for summer shade.

Evergreen trees are effective for wind protection during the whole year, but are not particularly good for shading. The way in which you can take full advantage of vegetation in the design of your building is taken up later in Chapter 7.

There are other concerns not directly affecting energy savings but which do affect your overall cost and your ability to do things with your site.

Depth and quality of soils (for growing things) are among these. So are the slope and moisture characteristics of soils for drainage, (for growing things and for constructing buildings and for site improvements); the presence of water (for amenity or as creating difficult conditions). Local U.S. Soil Conservation Offices can be a big help in providing this data on your site.

The data cited above will help you in selecting a general locality or specific kind of site. It also should add understanding to how the weather is affecting a site where you presently are living. For what you can do about it, go on to the next chapter.

SUMMER WINTER

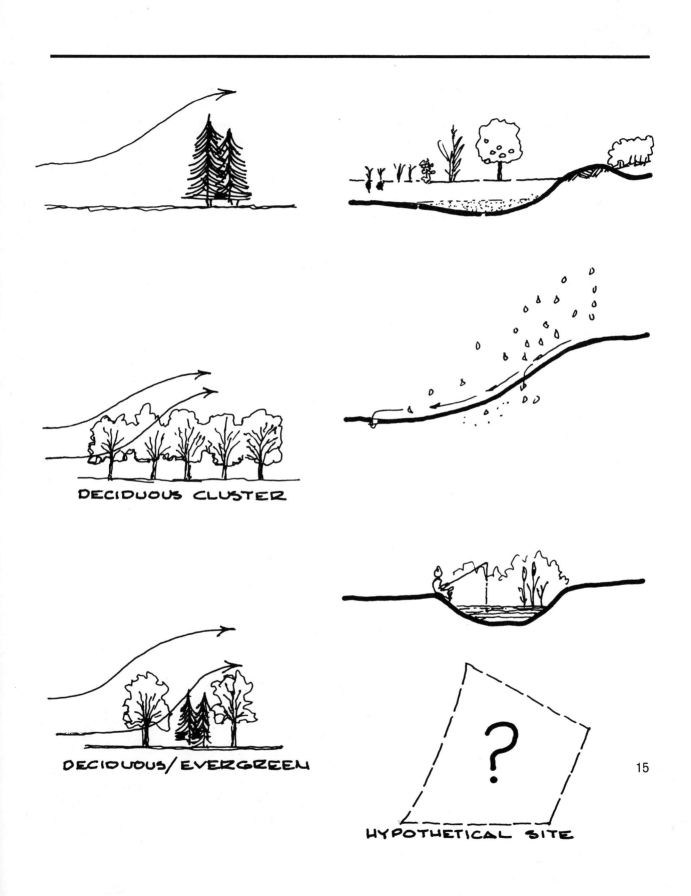

DECIDUOUS CLUSTER

DECIDUOUS/EVERGREEN

HYPOTHETICAL SITE

15

3

SITE PLANNING

By now you know your site intimately and are ready to start thinking about what you might do with it. Remember these two points, though, as you begin:

1. Energy saving is not confined to a building alone. The way you use your land or arrange the buildings and make improvements will save you energy over the long term and money during the building stages.

2. These same decisions made intelligently also can enhance the life and living pattern that you desire.

If not fully considered, these things may well cause all manner of problems further on in your project.

Take advantage of your freedom to start right from the beginning. Make full use of the fact that for once you are in charge of the design of this house.

Even if you are renovating an old place, this is a chance to take stock of the land that you have and make further use of it to emphasize and improve the things you enjoy. It is also the time to acknowledge and make the best of the difficult features.

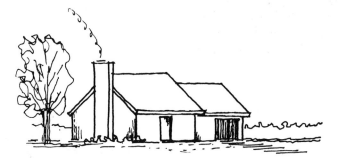

At this stage in your program you can make great savings in energy and dollars by planning to make sure of the natural elements — the sun and wind — to satisfy heating and ventilation requirements.

In addition you may make use of site features to protect it from adverse natural effects, and this also will lessen otherwise necessary expenditures.

Plan to provide for many of your living and leisure needs on your own land. Consider that this is more than just the usual house and garage. Think about growing foods. Think of the animals you may wish to keep, the leisure time activities or outside recreation you desire. Consider how your new home and home site may encompass them, too.

SUMMER

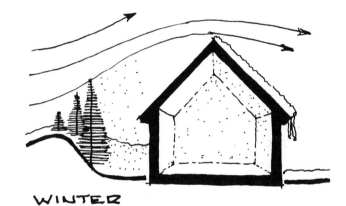

WINTER

SUMMER

WINTER

17

EXTERNAL FACTORS

The first considerations in your energy-efficient site plan are those external things that you can't change and that you should make maximum advantage of. These include:

Sun
Wind
Vehicular & Pedestrian Access
Utilities Access

Coming up here are factors which may be considered amenities: views, adjacent bodies of water, attractive nearby activities. Also consider the undesirable things — perhaps an unsightly view (to be avoided or screened), unappealing neighbors, sources of noise or nuisance.

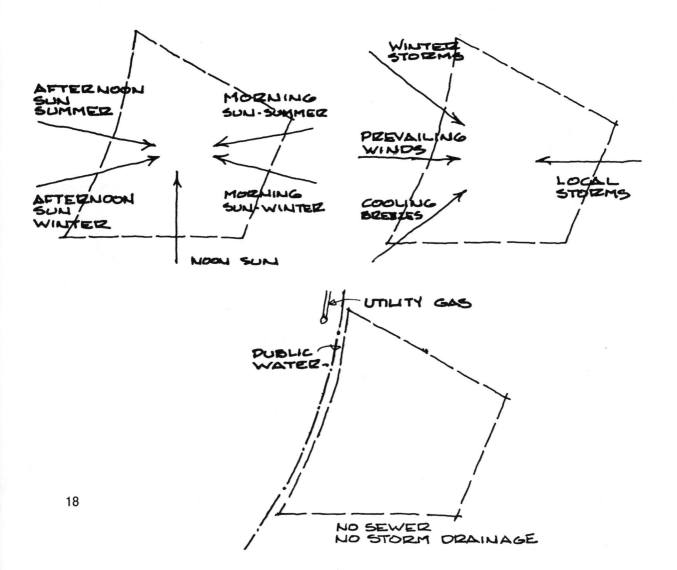

Make an inventory of the things here that you like and don't like about the site. Note those you can change, would like to change and must change. Include data on the topography or natural elevations and slopes of the site, soils, ledge or rock conditions, natural drainage patterns, existing plants and vegetations. Consider also natural features such as hilltops, rock outcroppings, streams, pools or wetlands.

SITE CHARACTER -ISTICS

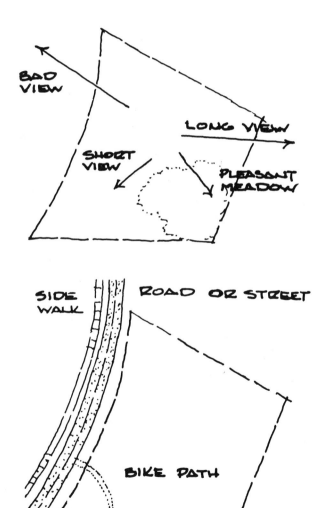

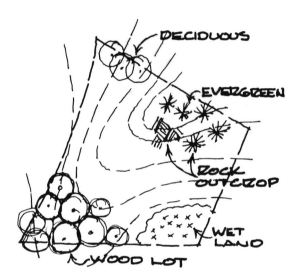

WHAT YOU WANT TO DO

The final step is to make a summary of what you wish to do on your new or recycled site. It may be a written outline, a shopping list or an expanded diagram. In trade talk this is called the "program."

The categories to note in your summary may include:

Shelter (this is assumed to be the major purpose).

Access & Services (both vehicular & pedestrian).

Growing Food — kitchen gardens, large family gardens, fruits & berries, orchards.

Animals — Pets, Useful animals — dairy, rabbits, pigs, sheep.

Recreation — tennis, swimming, flowers.

Business or Industry — cabin crafts, home industry, workshops, business or profession.

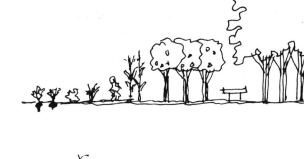

SUMMER

21

ANALYSIS

Now draft a series of diagrams (each of which illustrates the most desirable way of using the site for a particular aspect or purpose) like this:

Sun — for shade
Sun — for heat
Wind — for shelter
Wind — for ventilation
Use of Vegetation — for sun shading, for screening
Vehicular Access — maximum
Vehicular Access — minimum
Pedestrian Use
Utility Connections

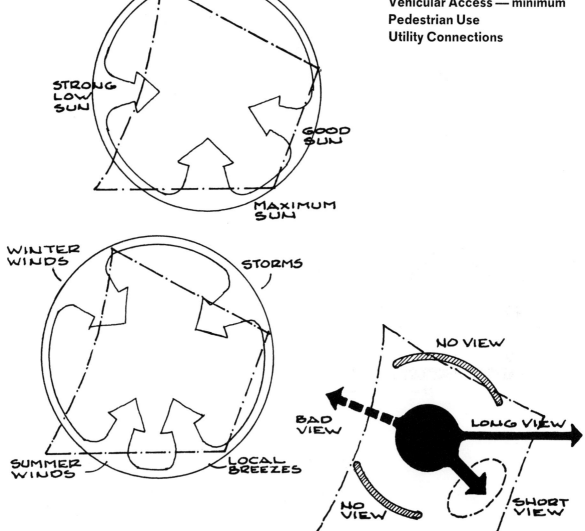

STRONG LOW SUN

GOOD SUN

MAXIMUM SUN

WINTER WINDS

STORMS

NO VIEW

BAD VIEW

LONG VIEW

SUMMER WINDS

LOCAL BREEZES

NO VIEW

SHORT VIEW

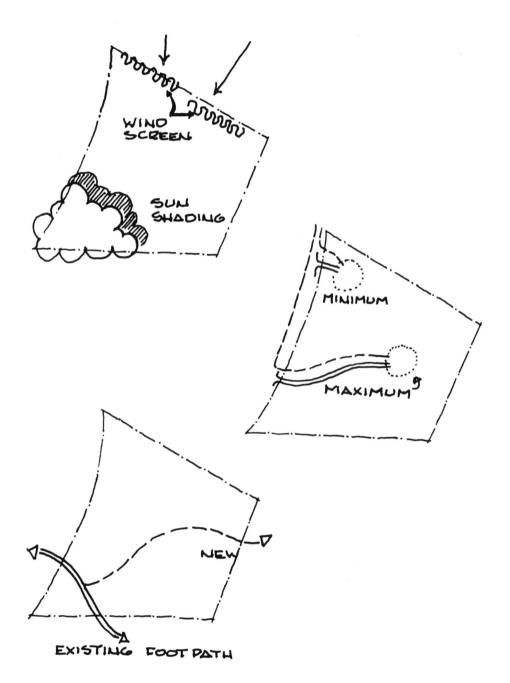

WIND
SCREEN

SUN
SHADING

MINIMUM

MAXIMUM

NEW

EXISTING FOOT PATH

DIAGRAMS

Use of Land:
    For gardening
    For orchards & diversified
    crops
    For animals
    For outdoor living
    For recreation pastimes

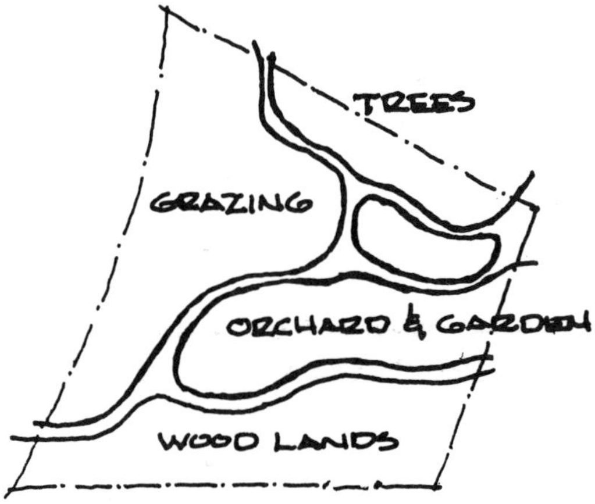

TREES

GRAZING

ORCHARD & GARDEN

WOOD LANDS

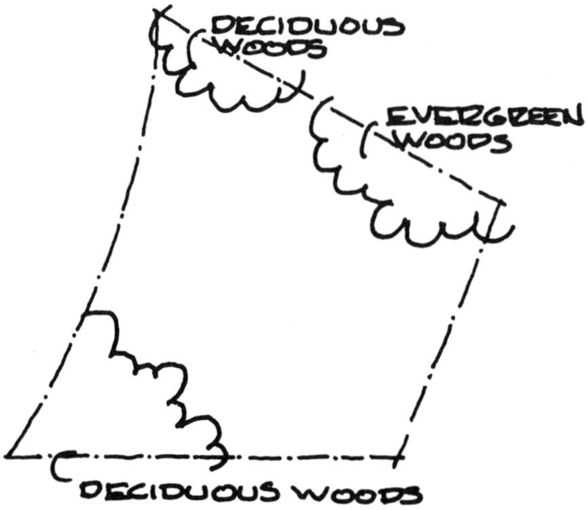

DECIDUOUS WOODS

EVERGREEN WOODS

DECIDUOUS WOODS

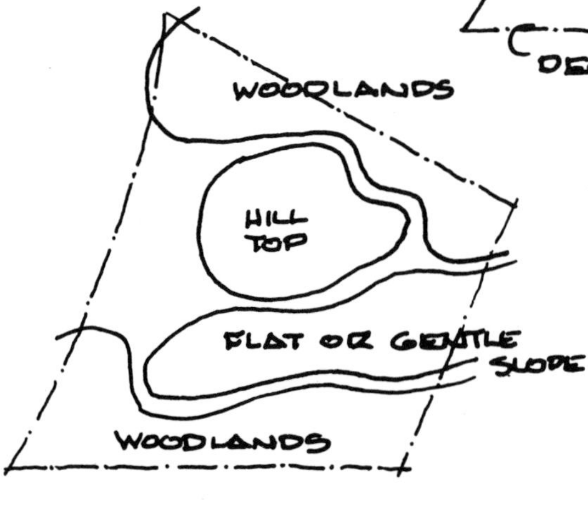

WOODLANDS

HILL TOP

FLAT OR GENTLE SLOPE

WOODLANDS

Draw up sketches also that
will show:

Privacy
Acoustic Isolation
Maximum Effect:
    Showplace (if desired)
    Inconspicuousness (if desired)

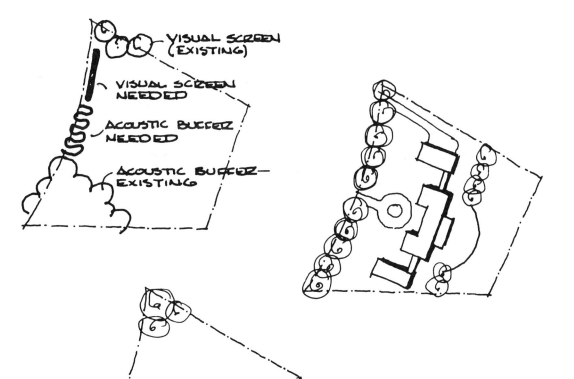

SYNTHESIS

The next step is to merge all of your diagrams. If you have thought ahead and drawn all your diagrams to scale on tracing paper or acetate film, you can combine them most simply by overlaying — that is laying one over another. Use different colors for each so that you can distinguish the lines and areas that you have drawn.

If you combine all aspects of all diagrams, it will be total confusion (unless you have been very lucky). So discard the diagrams or parts of diagrams which are of lesser importance or are obviously inconsistent.

Make modifications based on the opportunities you have perceived and recombine the rest, adjusting for obvious inconsistencies.

You should expect some conflicts. These will have to be resolved by your own judgement as to what is desirable and what is physically possible, all measured against potential costs.

Avoid what appears to be leftover areas or features. Try to assign either a positive (pleasant or useful) value to everything. Note all negative factors as a reminder to do something about them further along.

If there are many decidedly unpleasant or useless features or areas you may be looking at it from the wrong point of view. Even worse, you may have chosen the wrong site. Consider this very thoroughly now.

If you are dealing with a site that has existing buildings, they should be included as one of the diagrams and be combined with the others. They obviously will have an effect on the completed scheme.

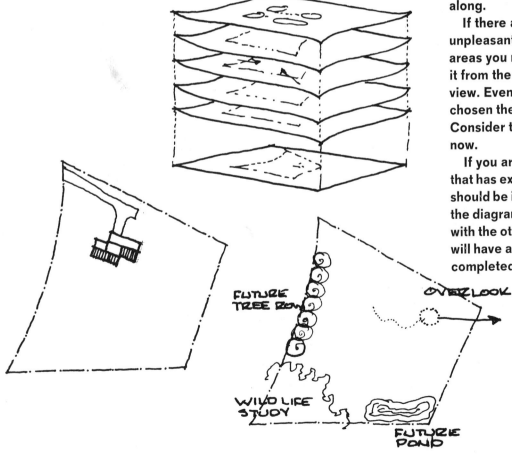

You may want to test your ideas and confirm your conclusions with a full size diagram on the actual site, using string, markers and stakes. Use helium filled colored balloons on strings to test elevations or make observations on hilly or neglected sites.

Make another composite diagram which may look like this:

Nothing replaces this first-hand, on-the-site experience, especially if you lack drawing skills.

If you had started out sketching in the normal way you might have drawn something that looked like this: Note how your composite plan differs, and move on to the next Chapter.

TRY IT OUT

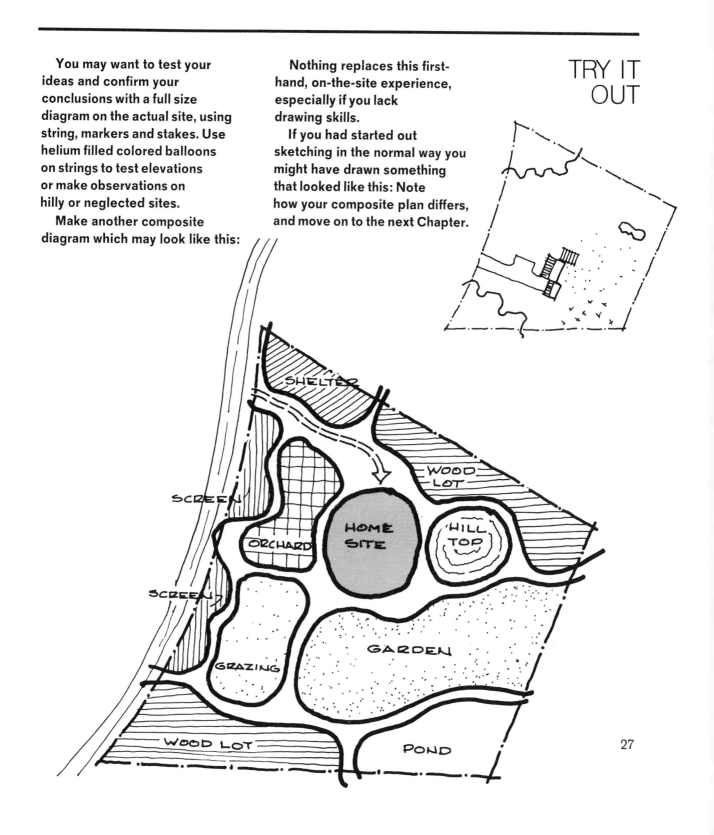

SHELTER

WOOD LOT

SCREEN

ORCHARD

HOME SITE

HILL TOP

SCREEN

GARDEN

GRAZING

WOOD LOT

POND

4

PLANNING THE BUILDING

You begin by thinking about what you want to do with the space inside your home or building. Think about what you do now, or would like to do, which you cannot do where you now live.

Consider individual and family needs, desires. What are the known activities? Consider the needs to have quiet, to read, study, write, and also the desire to work at something perhaps noisily, or messily; the habit of children to get together and do the things they do which are important; or the needs of a child to be alone sometimes, which also is important.

Consider spaces in terms of the activities they will contain, call them food preparation, eating, (or dining, for those where there is a difference), studying, sleeping, washing, sitting, hobbies & crafts, sewing, entertaining, family gathering, storing, laundering.

Avoid thinking in specific traditional terms as "living room." Everybody in fact lives, and how they do it has nothing to do with 12 X 20-foot dimensions that are the most striking characteristics of such rooms. Forget about "dining room," "kitchen," "family room."

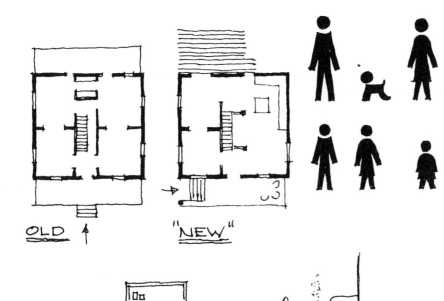

OLD

"NEW"

The food preparation space, usually called the kitchen, by far is the most intensively-used space in your home. It is often the least-considered and yet the most overdone space, usually emerging as a collection of shiny appliances and ersatz wood cabinets jammed into odd shaped and sized space. The food preparation space, if nothing else, should express the individual lifestyle of the home owners.

For some families it can be a large, pleasant space for preparing and consuming food, also a place for many household tasks, a place where children play and adults may entertain. This use of the space, the heart of the household, usually is identified as a farm kitchen.

At the other extreme, if the need is for a small compact space where simple meals are prepared to be eaten elsewhere, a galley kitchen will suffice. A gourmet cook will want space which approximates a studio in the care with which it is laid out and appointed.

The family that preserves a great deal of its food needs a place for all the proper utensils, with ample space to use and store them. Often this is identified as a harvest kitchen.

Think of such a space in your own terms. Then plan, build and enjoy it afterwards.

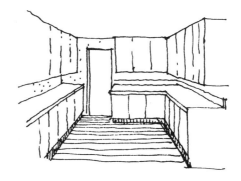

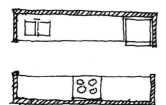

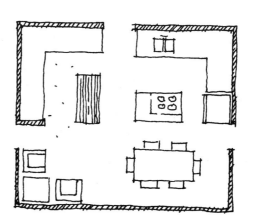

29

STORAGE
–FOOD &
OTHERWISE

Most houses come with built-in storage in the form of closets for general storage and wall and base cabinets for more specific items such as for food and utensils.

Cabinets tend to be inflexible and expensive. Consider instead open shelving where what you store is easily visible and can add decorative appeal. Think of large bin-type storage too, which is good for large volumes of foodstuff that are least expensive to buy. What about a closet with floor-to-ceiling shelves, or a separate room with all walls of shelves, racks or bins? Consider that a whole new cash saving food purchasing program can be generated if there is proper storage for it.

Such storage walls are good in other parts of the home and can be used as moveable room dividers or as improved accoustic barriers.

Keep in mind that the most efficient storage is where the whole area is easily available. Avoid deep closets except for real "dead" storage.

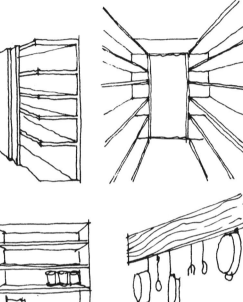

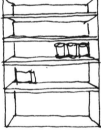

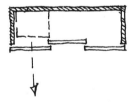

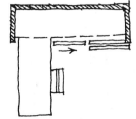

In thinking out your needs for other types of space, be thorough in examining the main way it will be used, the number of people who will use it and the variety of activities that may go on there.

Consider how listed needs should be accommodated — whether large open areas for a variety of people or activity, or a series of small, specifically-designed spaces.

Keep in mind that some activities demand a character of space which prevent its use for anything else. People desiring a dining area for intimate dinners must keep in mind that such a space is good for little else except when planned as part of circulation.

Open plans such as seen in primitive house and in traditional farm kitchens, as well as contemporary houses, have a great deal to say for themselves in efficiency and flexibility.

The opposite — closed plans, I suppose — where each well-defined, enclosed room has a specific function emphasizes privacy and the individual character of each space.

USE, QUALITY & CHARACTER OF SPACES

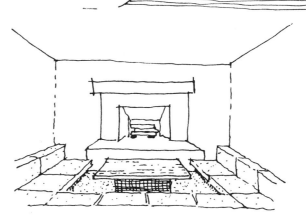

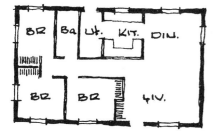

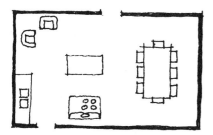

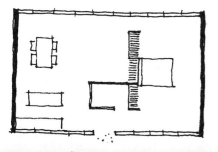

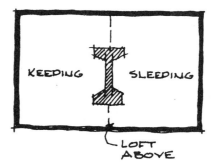

KEEDING   SLEEDING

LOFT ABOVE

GETTING
ABOUT

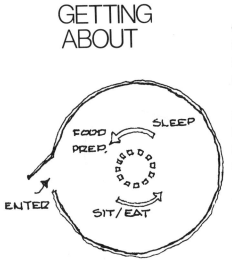

In addition to assessing the character of individual spaces, consider how they may relate to each other, to the outside, to specific site elements.

A major consideration in planning any building is how you move from one activity or space to another. Call it circulation.

In a simple one-room structure that contains a number of activities, you simply shift from one activity to another. Except for the furnishings and fixtures, the space required for circulation is not at all defined and generally is minimal.

In a more complex building with individual rooms for individual activities, the space required for getting about is greater, and it becomes well defined and expensive.

The important nodes of circulation are where you enter from outside, where you change levels, and, obviously, areas of high family use such as bathrooms.

It's a good idea to start analyzing circulation in your existing home. See if it works efficiently for your needs. An old house may be greatly improved merely by simplifying or defining the circulation. A new front or back door or one or two interior doors may do it. On the other hand, in older homes the large halls and grand staircases once intended for circulation now may lend themselves to new use.

Don't forget the outside, too. Interior circulation should hook onto exterior, and this relates also to site planning.

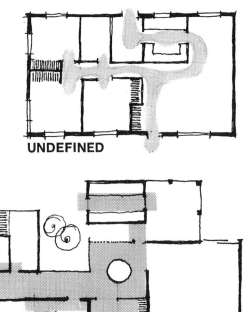

UNDEFINED

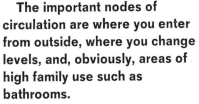

EXISTING

DEFINED

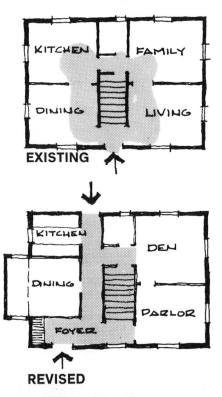

REVISED

By now you should be making notes or lists of the kinds of spaces which you need. They should have names or descriptive identifications and a a word or two about each activity undertaken there. For a small family it may look like this:

ACTIVITY   TYPE SPACE

PREPARE FOOD ⎫ FARM KITCHEN
CONSUME FOOD ⎬
FAMILY SOCIAL ⎭

CHILDREN PLAY ⎫ CHILD'S BEDROOM
SLEEP-CHILDREN ⎭

SLEEP- ADULT...BEDROOM
ACTIVITIES-ADULT-STUDY- SITTING RM
HOBBIES .... WORKSHOP
STORAGE .... LOFT/ATTIC/GARAGE

OR

PRESERVE FOOD ⎫ HARVEST KITCHEN
PREPARE FOOD ⎭

SERVE FOOD ..... DINING
ENTERTAIN ...... PARLOR, "GREAT HALL"
ADULT ACTIVITIES-STUDY
CHILDREN PLAY ⎫ DEN / FAMILY RM.
ENTERTAIN ⎭

SLEEP-------........ BED ROOMS
STORAGE ...... CLOSETS/
                STORAGE WALLS

Note that both lists include a wide variety of activities and kinds of spaces.

Now try to group together the spaces which have an affinity of use of physical need.

This is your program. To make it more useable you might add your guess as to the required dimensions of the spaces sought, and some notes as to their locations, orientation or character.

For an individual with many diversified interests and who entertains occasional guests, it might be thus:

PREPARE FOOD...GALLEY
SERVE MEALS....DINING ROOM
ENTERTAIN ......-DEN/CONVERSATION
                    PIT
QUIET ACTIVITIES-LIBRARY
ACTIVE   "   .WORKSHOP
SLEEP .... BEDROOM
STUDY/WORK...STUDIO
STORAGE ...... "OUTFITTING" ROOM

DIAGRAMS – GENERAL

Please don't start drawing floor plans with square boxes for rooms. Take a few more minutes and develop thought diagrams of what you want to do with the spaces.

At this stage you have in mind that:

You have to get in or out of the house, perhaps at more than one location. You have to get around inside. You should group together certain uses, particularly those that need pipes to serve sink, baths, lavatories, etc. Some spaces may need to be isolated for quiet or privacy. Some may want to be convenient.

Consider again that there may be a view to be taken advantage of (or avoided) or need for orientation to sun or toward (or away from) winds.

Draw your diagram with a series of labelled circles. Use lines — both full and broken — to show circulation between them, arrows to show important directions or exterior influences.

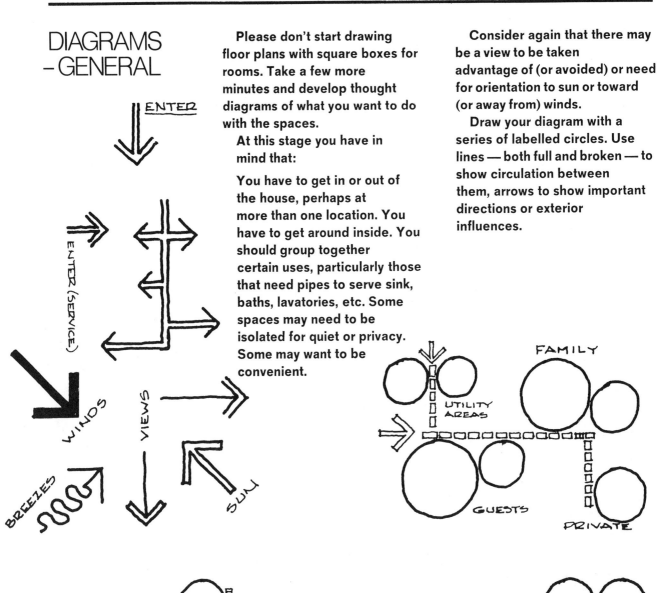

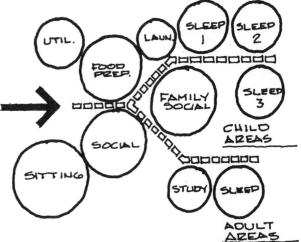

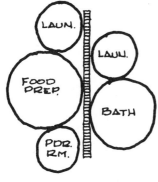

For example, the family which sees itself as close-knit and spending much time together, the diagram may look like this:

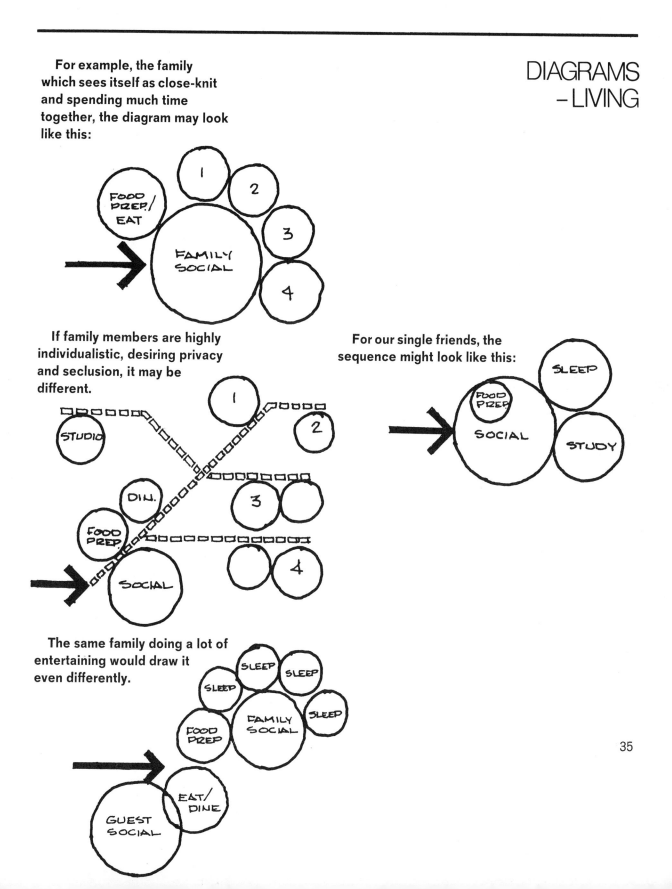

If family members are highly individualistic, desiring privacy and seclusion, it may be different.

For our single friends, the sequence might look like this:

The same family doing a lot of entertaining would draw it even differently.

35

ORIENTATION OF SPACES –SUN

Orienting major-use rooms to the southern or warmer exposure, makes maximum use of the solar energy.

An eastern exposure is second best, making the most of the morning sun. Southwestern exposures are adequate.

Western exposures present the difficulty in controlling low-lying late afternoon sun, and this also is often the source of prevailing winds.

Northern exposures receive no direct solar radiation and therefore should be consigned to infrequently used spaces and those which require little or no heat. Spaces that are heated to a very low level or not at all offer very good insulating layers for these exposures.

If maximum shading from the sun is desired, the opposite from the above would apply, with easterly and northerly exposures being favored.

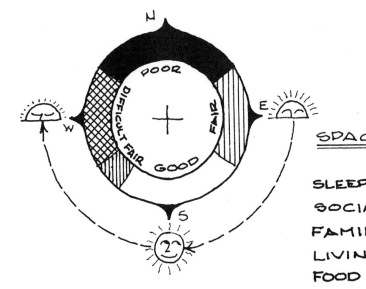

SPACE	N	NE	E	SE	S	SW	W	NW
SLEEPING	O	O	O	O	O	O		
SOCIAL				O	O	O		
FAMILY					O	O	O	O
LIVING					O	O	O	O
FOOD PREP.	O	O	O	O				
PLAY SPACE			O	O	O			
STUDY	O	O	O	O	O	O	O	O
WORK/HOBBY	O	O	O	O	O	O	O	O
BREAKFAST		O	O					
LUNCH						O	O	O
DINNER					O	O	O	O
STORAGE	O	O					O	O

Building surfaces that stand 45 to 90 degrees off the direction of the wind, offer the least wind velocity on the surface and least heat loss. Building openings made for natural ventilation are best when placed at right angles to cooling breezes. If cold winter wind and summer cooling breezes are from the same direction, the following arrangement is suggested:

Ideally, the direction of desirable and undesirable winds would vary by 60 to 90 degrees, which would allow the best of both situations.

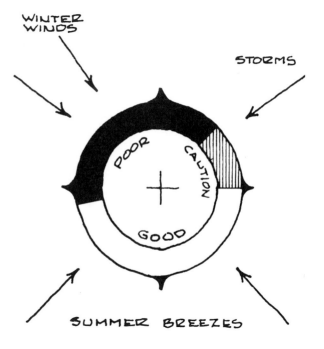

WINTER WINDS

STORMS

POOR

CAUTION

GOOD

SUMMER BREEZES

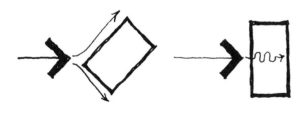

COLD WINDS

SUMMER BREEZES

DIAGRAMS
–ENERGY

Now add the sun and wind considerations to your living diagrams. Also incorporate the outdoor and site features.

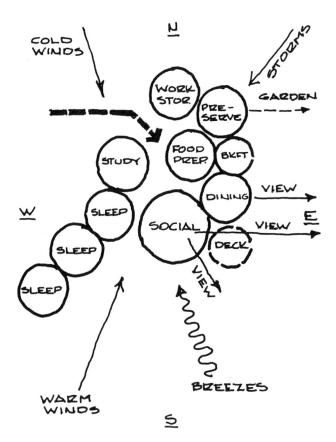

One of the first concrete decisions you can make on the energy-saving characteristics and the appearance of your house will be its shape, or its three-dimensional volume. There are two basic approaches.

The first approach is to take a preconceived form—either because you like it or it is available — and divide it up. This adapts to simple program needs and uncomplicated inside spaces.

Such simple geometric shapes as circles, octagons and the three-dimensional cube adapt best to this approach.

Domes, yurts and A-frames are well-known building types that also work particularly well in this approach. But when there are more complicated needs, dividing or adding to forms become complex, and room shapes are difficult.

The second approach is suitable for more complex plans and it is more responsive to climatic conditions and energy savings.

Here you try to devise a single shape, simple or complicated, or a series of shapes which fulfill your needs, which are appropriate for the climate and which will assist in saving energy.

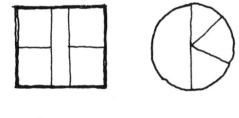

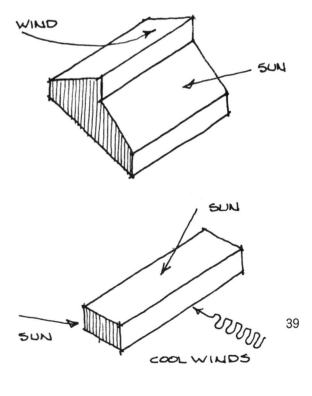

WIND

SUN

SUN

SUN

COOL WINDS

39

FORMS
& VOLUMES
– A DISCUSSION

Here are a few general rules: A cube encloses high volume in relation to its surface area and thus gives limited opportunity for heat loss in cold climates, or excessive heat gain in hot ones. A sphere, part of a sphere or any circular form, is still better, although somewhat less useable on the inside.

Among the more ordinary forms, compact, rectangular shapes which approach a cube are the most efficient for a building or parts of a building complex. Multiple-story buildings are more efficient in terms both of retaining heat and having the least area available to the sun for heat gain or heat loss.

Long, rectangular or linear shapes are best for offering maximum surface to the winds for cooling.

Rectangles with proportions in the range of 1:1.1 to 1:1.3 are considered good compromises for both solar and wind orientation.

The minimum wall area and the large pitch roof areas are faced toward the cold winds.

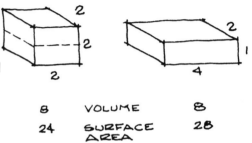

	VOLUME	
8		8
24	SURFACE AREA	28

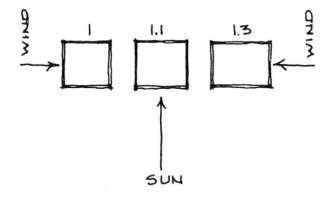

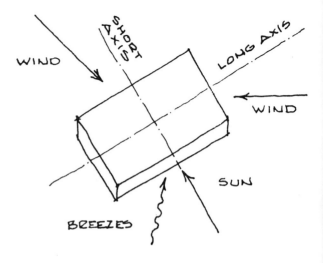

Considering climate and enclosure factors, we can draw typical building volumes that are appropriate for varying climates.

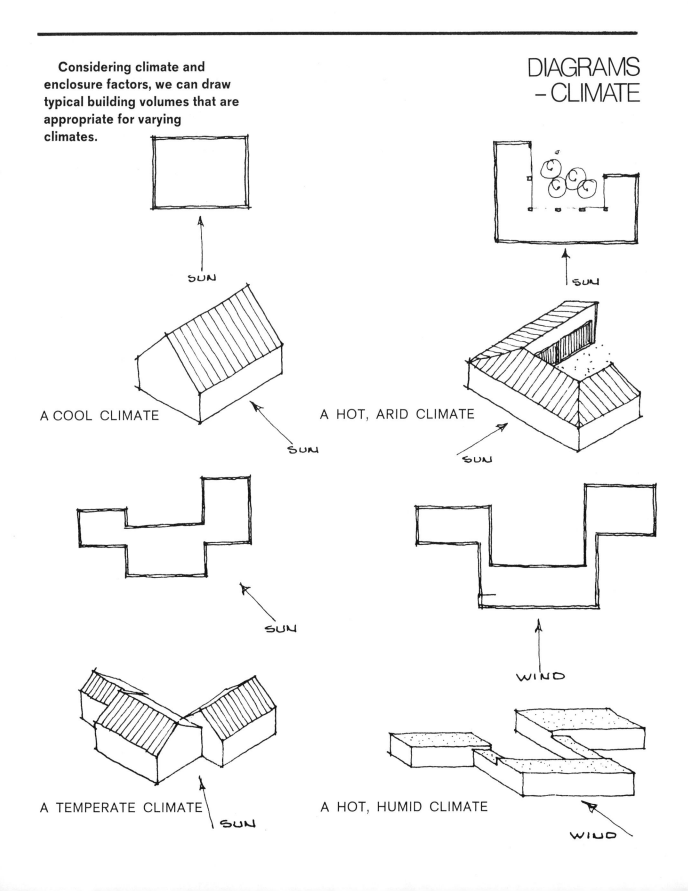

SUN

SUN

A COOL CLIMATE

SUN

SUN

A HOT, ARID CLIMATE

SUN

SUN

WIND

A TEMPERATE CLIMATE

SUN

A HOT, HUMID CLIMATE

WIND

HOW MANY BUILDINGS? HOW MANY FLOORS?

Good clues are obtained from your composite diagram as to whether your building should be spread out or compact or have more than one floor.

If you desire a tight cluster of closely-related activities, a single form — perhaps with a single space—is indicated.

A series of activities along a single "spine" indicates a linear organization, perhaps like a ranch house.

A series of clusters of activities having some relationship to one another would lend itself to a number of levels.

Clusters of activities having no relationship might indicate a whole series of buildings or structures.

Think of your diagram also in relation to time. Perhaps you don't wish to or can't afford to build everything at once. Assign priorities to your desired activities and work them into your diagrams to build them over an extended period. But try to maintain the integrity of your desired diagram throughout.

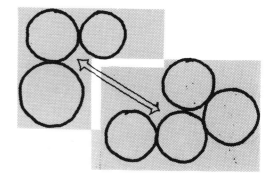

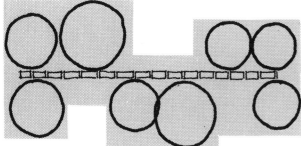

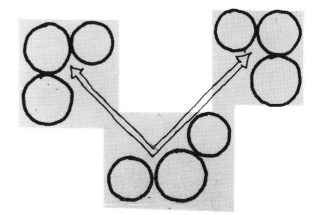

If the goal is renovation or adding to an existing home, the process is the same, although somewhat more challenging. The task will be to make the same kind of program and draw the same desired diagrams as described earlier. Even if you have a good idea of what you want, a list and a series of diagrams will help you sort things out and get started.

Such an effort may be helpful in determining the value of your present place for renovation for your use. It may even convince you that you should start all over again. On the other hand, it might help to show many of the hidden possibilities of the old place.

After working up the program and diagrams, try to fit the diagram within your existing volume and spaces.

If most of your needs are contained within the existing volume, with only occasional bulge or modest appendage, reassigning room uses and adding some space is the most desirable course.

When a diagram shows a large number of added activities and spaces, but which may relate to existing areas, perhaps an extension is called for. Clusters of activities not particularly related would indicate whole new building masses.

ACTIVITY	HAVE	DESIRE
PRESERVE FOOD	....	→ ○
PREPARE FOOD	KITCHEN	
STORE FOOD	....	PANTRY
SERVE FOOD – FAMILY	....	{ BRKFST RM. FAMILY RM.
SERVE FOOD – SOCIAL	DINING ROOM	
FAMILY – SOCIAL	....	DEN/FAMILY ROOM
GUEST SOCIAL	LIVING RM.	
CHILD SOCIAL	....	LOFT OR → PLAY ROOM
CHILDREN PLAY	{	
SLEEP	.... → BEDROOMS →	
HOBBIES	...... WORKSHOP	
STORAGE	..... ATTIC GARAGE	CLOSETS/ WORK WALLS

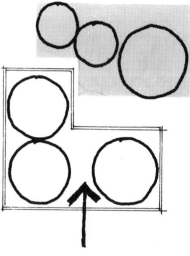

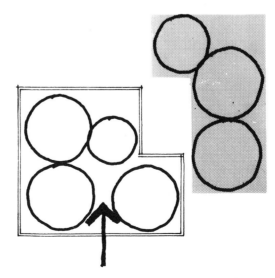

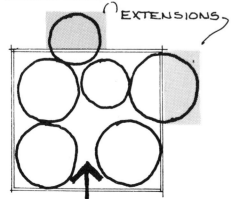

EXTENSIONS

HOW DO YOU ADD TO EXISTING?

This section consists of a few tricks for the person who already has a building and wants to add to it.

Increase existing volume by:

Adding: Lean-tos
Bay windows
Sheds
Repetitive forms
larger
smaller

SHED & LEAN TO

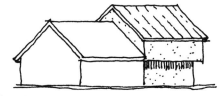

REPETITIVE FORM - HIGHER

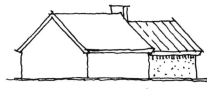

REPETITIVE FORM - LOWER

Extension:

New building:

EXTENSIONS/ BAY WINDOWS

NEW BUILDINGS

5

PUTTING THE HOUSE TOGETHER

We now come to the ways of enclosing your emerging energy–efficient house, and we will consider the house in three parts: the part that shelters you from the elements — the roof or cover; the part that sits on the ground; and that in between which joins the two — which defines the space where you generally move about and live.

The roof and foundation, while often thought of in purely functional terms, can have real influences on what goes on in your house. They also can affect your use and expenditures for energy.

The sides of a house, too, can be more than walls with door and window openings cut into them. The arrangement and design of these elements can provide energy-saving as well as creating exciting consequences in how you live.

JOINING THE BUILDING TO THE LAND

You doubtless will require some way to connect your structure to the ground. The reason is primarily for support — in some localities to prevent it from blowing or floating away — but if properly considered, underpinning also can significantly improve upon your living patterns and energy consumption.

Footings and foundations consist of isolated — usually masonry or concrete — piers, or ordinary timber or metal posts; Continuous — wall type foundations with either a crawl space or fully excavated basement; or Spread (basically a flat slab) upon which a structure is placed. The last is usually constructed of concrete but may be layers of any rigid material such as wood. In climates where the ground freezes in winter, the bearing section of the footings must extend to below the maximum penetration of freezing or "frost."

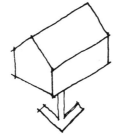

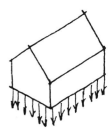

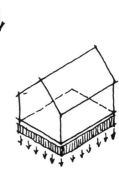

You should consider the following pros and cons about the various types of foundations:

1. Isolated piers and post footings are the simplest and usually the cheapest. The underside of the floor surface, being exposed to cold, must be insulated. On the other hand in warm climates, this provides ventilation and separation from ground moisture.

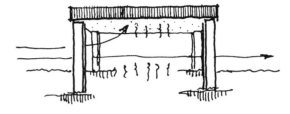

2. Continuous footings and foundation walls probably are the most expensive and complicated. They do yield a "warm floor" and even allow the building to receive heat from the earth. They also provide in the basement a place for a central heating system (almost ideal for a simple gravity air system because the heat rises). They also provide additional working and storage space.

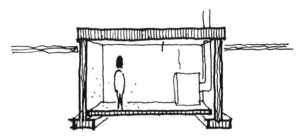

3. Spread arrangements are simple and quick to build. In cold climates they require a deep layer of well-drained soils to act as support and insulation under the slab or vertically on the edge or frost walls. Spread foundations are best on flat sites and well-drained soils.

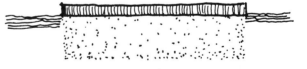

47

INSULATION

FROST WALL

FOUNDATIONS & THE SITE

The type of foundations should be chosen according to your site. On sloped sites and areas of rock or ledge, individual piers or posts are simplest, there being little or no excavation required.

Continuous foundation walls with basement are reasonable if excavation is possible — even desirable if you can use the slope to provide a bonus of enclosed space.

On sloped sites slabs or spread footings are a disaster. Unless you can document a special need, don't even consider one.

Bear in mind that on flat ground post or piers will be above the ground (for ventilation). A continuous wall foundation will yield only "cellar" type spaces. Trying to put major rooms or activities in such a basement is not really worth it.

On a flat site the slab approach really works well for you. The house fits nicely on the ground and construction is simple.

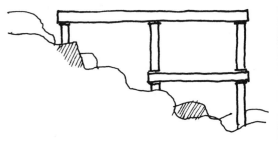

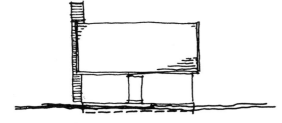

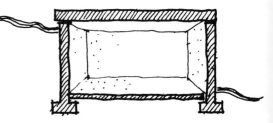

Some of the advantages to
consider when thinking about
foundations, then, are:
    Variety of levels
    Quality of spaces
    Heat from the earth
    A total-foundation house

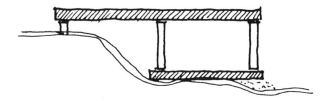

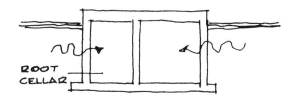

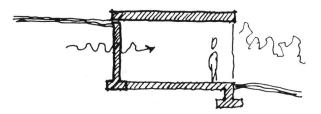

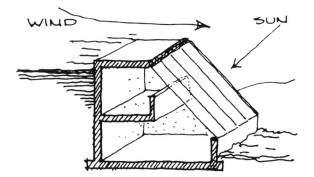

THE ROOF OR COVER

The roof of your house is more important than most people think — more than just a cap or a lid. Primarily it is shelter, and is the reason for enclosing space.

It is shelter from rain in all climates, snow in colder ones, and sun in warmer ones. It also is shelter from wind and windblown particles. The exterior surfaces and overall shape will determine how effective it is as a shelter and will largely determine what your house will look like. Its underside will determine the quality and character of the space you are going to live in. Next to joining your house to the earth, perhaps the most important thing is how you relate it to the sky.

Roofs are either flat or pitched. Flat ones are useful for catching and retaining things like rain in arid climates, the sun's rays on solar panels or snow for added insulation in cold climates. The space also may be useable for decks or gardens.

But since flat roofs do not readily perform the first requirement of a roof — that of shedding water — unless you have a compelling use such as those listed above, we suggest you don't consider one.

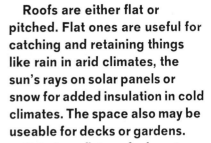

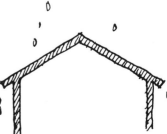

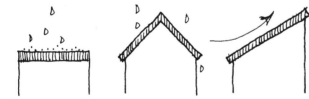

Pitched roofs which shed water best, can be characterized as low-pitch or steep-pitch. Those steep-pitched are good for shedding snow as well, and they also enclose space to become a diagonal wall system.

Structures like tepees, domes, yurts and A-frames can be thought of as "total" roofs. The same surface forms both roof and walls.

A low-pitched roof (between 1:12 and 6:12) will shed water, and the lower pitches will retain snow for insulation. The inside area will have extra space and pleasantly-sloped ceilings.

A steep pitched roof (6:12 to 12:12 and above) will shed almost everything. That's the reason for the pitch. Since it requires more material to form the larger surface, it is more expensive. You might as well benefit by making use of the extra space, but avoid inaccessible, unpleasant cramped, attic rooms. They may be great for storage, though.

Steep pitches also are good for placing solar collectors in northern climates.

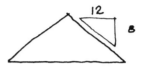

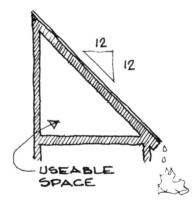

USEABLE SPACE

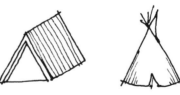

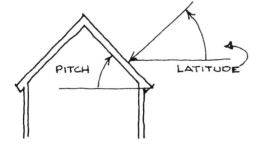

PITCH   LATITUDE

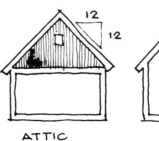

ATTIC

LOFT

51

STRUCTURE

A note on structure-bearing systems — There are two basic types:

1. In a frame-bearing system all loads are transferred to main beams and girders, and then to columns and to the ground. In this case added "skin" or "infill" is needed to actually enclose the space and keep out the weather.

2. In a wall-bearing system the total length of the wall supports the loads and carries them to the foundation. In this approach the walls perform the dual function of support and enclosure.

Ordinary wood framing, either platform or balloon, combines the two. Continuous wood studs of the enclosing walls and partitions support both the loads above and the skin of the building.

There are, in addition, domes, space frames, stressed skin structures and others, but most can be thought of in terms of the dual use of providing support and enclosing space.

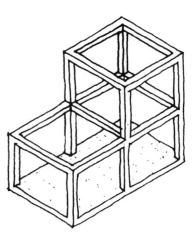

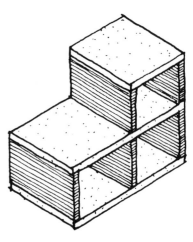

A major loss of energy in cold climates comes through the outside of walls. Heat is lost by conduction through the wall surfaces and by actual infiltration of cold air through cracks and openings.

A related phenomenon, which also contributes to energy consumption, is the build-up of cold air on the inner surfaces, particularly windows. The air then cascades down the surface, causing drafts.

Insulation, in the form of trapped air, will prevent heat conduction from the inside out and cold from outside in. Inside drafts and areas of localized discomfort can be avoided by trapping cool air which has built up against cold surfaces.

Another aid to retaining heat is to provide a layer of stable air on the outer (cold) wall surface. This provides an additional insulating blanket.

It is the removal of this layer by wind action, together with the actual infiltration of air, that causes buildings to be markedly colder on windy days.

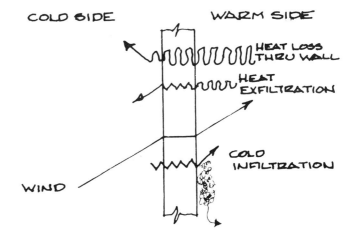

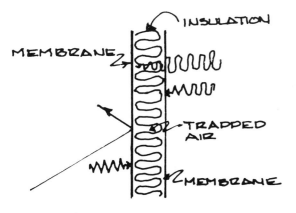

CONSTRUCTION OF WALLS – EXAMPLE

The design of walls aims to retain heat inside the structure and prevent cold air outside (in the form of infiltration) from entering the envelope of the building.

A secondary but almost as important consideration is preventing undue or unwanted solar heat gain in summer, which would require energy to dispel. Fortunately a well-designed wall will act both ways.

A wall well-designed for a comfortable, efficient space will have as many of these elements as possible:

1. A weather surface — to shed water and protect the rest of the wall elements.
2. An exterior membrane — to act as backup of weather surface to exclude water, to prevent infiltration of wind-driven moisture or air itself.
3. Structural solidity — to provide support and rigidity of the wall and other building elements.
4. Insulation — which generally consist of an entrapped volume of air, which prevents conduction of heat through the air to the outside.
5. Air space — to provide ventilation for other elements of the wall and act as increased insulation.
6. A vapor barrier — to prevent moisture in the form a vapor from the inside from entering the wall; also to help prevent infiltration of exterior air.
7. The interior surface — provides wearing surface for interior activities, assists in insulation, prevents infiltration and provides surface for decoration.

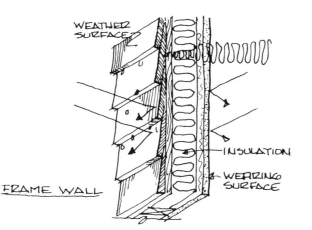

FRAME WALL

WEATHER SURFACE

INSULATION

WEARING SURFACE

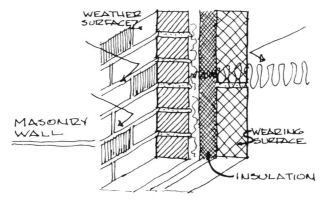

MASONRY WALL

WEATHER SURFACE

WEARING SURFACE

INSULATION

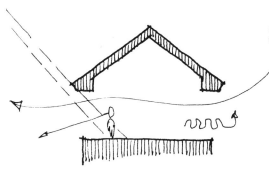

Between the joining to the ground and the ending at the sky is that area between where the walls occur. This also is where the doors, windows, a large amount of natural light and views are enjoyed, where natural ventilation happens, and where a great deal of heat loss (or gain) takes place.

To this area we also must consider how to attach the previously-mentioned roof to the previously-mentioned foundation. Also there is the support of any floors or spaces in between — a matter of structure.

We also should consider this area's protection from wind, noise and unsightly views.

Energy-saving factors at this point are to reduce use and to increase comfort by cutting down exposure to the severe elements, and to provide in-sulation to retain heat. Openings, especially, should be carefully planned to prevent heat loss and unwanted or unneeded heat gain. Openings also should allow for desirable heat gain and natural ventilation.

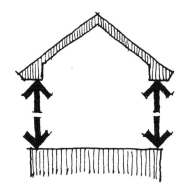

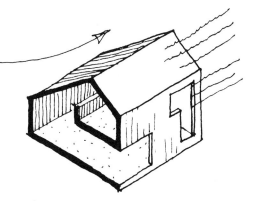

INSULATION

Insulation, the means by which cells of air are trapped in the wall, comes in several types. Those industrially produced are most available and most effective at present. They can be described as follows:

Batts and Blankets — prepared volumes of expanded glass fiber, mineral fiber or organic fiber which are placed in the wall, usually between wood or metal framing members.

Boards — rigid sheets of foamed cellular plastic or cemented together particles or fibers. Those are most efficient per unit of thickness and are good for adding to masonry or concrete surfaces as well as adding to surfaces of existing construction.

Blown or Poured — loose expanded mineral or organic fibers which are placed or blown into frame spaces. These are the most useful for insulating existing buildings.

Foamed-in-place — cellular plastic foam which is injected into framing spaces and then solidifies. This also is useful in insulating existing buildings.

Of the above, blown and foamed-in-place require a commercially available process with special equipment. The other types are simple to install and can be done by most anyone.

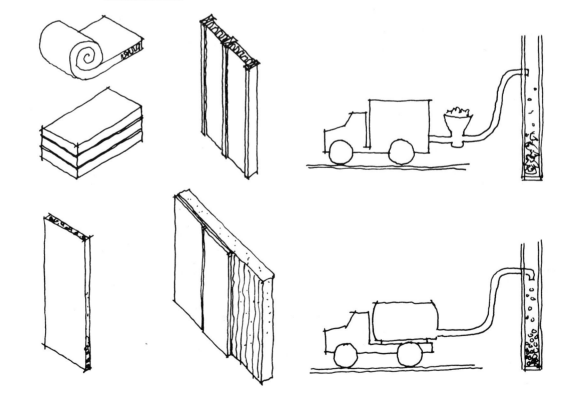

The loss of heat through existing window openings can be reduced by:

1. Adding an additional layer of glass (storm sash or panel). This is most desirable if only a single layer of glass is currently provided. If two layers are presently installed the addition of a third will increase the benefit.

2. Securely sealing cracks around glass, sash and window frames to prevent air infiltration.

3. Providing moveable closures or shutters on the exterior to trap an added layer of air next to the glass. This also will prevent "windscouring" of the surface. Such closures, in the form reminiscent of rolling "barn doors," are most effective on large, existing glass walls.

4. Installing similar closures or heavy drapes on the inside for the same effects.

5. Providing shelters for window openings in the form of overhangs, baffles, recess or plantings.

ONE LAYER

TWO LAYERS

THREE LAYERS

SEAL

SEAL

DRAPES

INSULATED SHUTTER

SHELTER

HEAT LOSS THROUGH DOORS

Heat loss through entrances may be reduced by many of the same means as windows. An added storm door is the first option, followed by weatherstripped and well-sealed frames.

Adding a vestibule to form an air lock not only will save heat, but will cut down uncomfortable drafts.

If doors are in an exposed location, overhangs, baffles or plantings can protect against the full force of the wind.

An effort to provide a sheltered place or orientation for doors and entrances will be rewarded with lower heat bills and a more comfortable home.

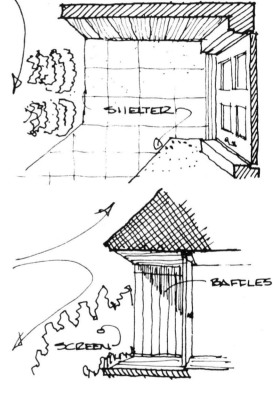

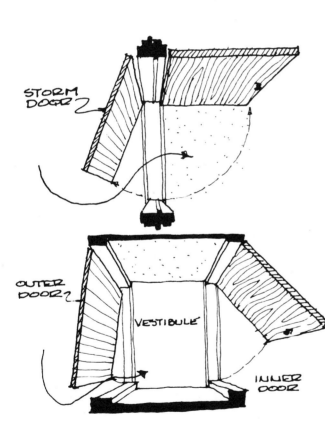

STORM DOOR?

OUTER DOOR?

VESTIBULE

INNER DOOR

SHELTER

BAFFLES

SCREEN

Your window openings are planned to provide natural light, in part for ventilation and also to provide a feeling of release from the enclosed space. The openings also will take advantage of views and establish a visual continuity with your outdoor areas.

Windows traditionally have caused major heat loss during cold periods. Recently-developed large glassed openings have increased the heat loss and also become a source of unwanted and uncontrolled solar heat gain during warmer months.

The glass openings of your energy-efficient house need not be a major energy drain. They can bring the benefit of solar heat and natural ventilation, thus reducing the need for and cost of energy for these operations.

WINDOW OPENINGS – FOR LIGHT, VENTILATION & VIEW

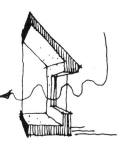

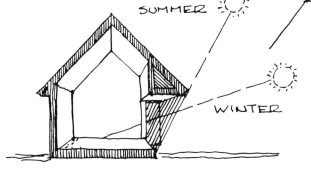

SUMMER

WINTER

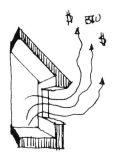

WINDOWS – LOCATION & SIZE

Windows facing east-south-east might be the largest, since these are the warmest exposures and will cause the least heat loss. Openings on the west should be small or carefully designed to conserve heat. Windows on the northwest and north should be avoided or made as small as possible.

Southern and southwestern windows have the best potential for controlling solar heat impact and, with proper design, they can make good use of solar heat during cold months. Western windows represent a difficulty in controlling the low lying sun. Window openings on the north obviously are unaffected.

Large openings allow the best possibility for natural ventilation when arranged to allow a cross-current of air. Openings should be oriented to pick up prevailing summer breezes.

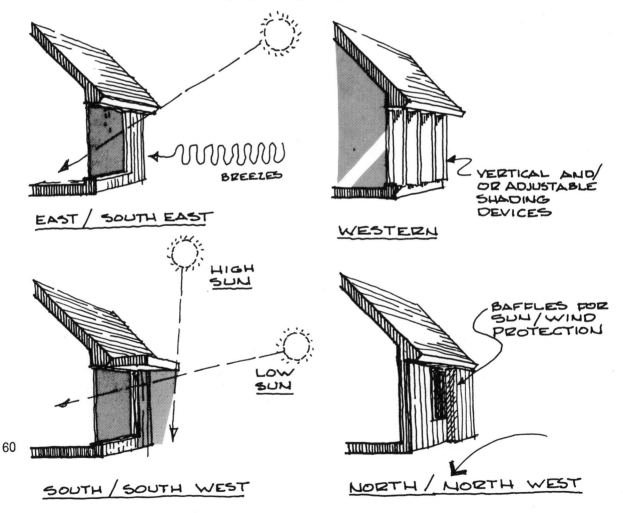

BREEZES

EAST / SOUTH EAST

VERTICAL AND/ OR ADJUSTABLE SHADING DEVICES

WESTERN

HIGH SUN

LOW SUN

SOUTH / SOUTH WEST

BAFFLES FOR SUN / WIND PROTECTION

NORTH / NORTH WEST

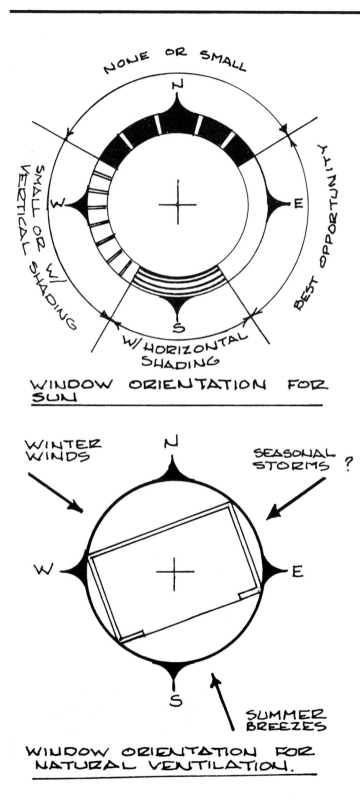

NONE OR SMALL

N

E

SMALL OR W/ VERTICAL SHADING

VERTICAL SHADING

W

BEST OPPORTUNITY

W/ HORIZONTAL SHADING

S

WINDOW ORIENTATION FOR SUN

WINTER WINDS

N

SEASONAL STORMS ?

W

E

S

SUMMER BREEZES

WINDOW ORIENTATION FOR NATURAL VENTILATION.

61

OPENINGS FOR NATURAL VENTILATION

Plan the openings to provide for cross-ventilation insofar as possible.

Good air flow occurs when the inlets and outlets are approximately the same size.

A larger area of outlet than of inlet will allow faster air flow and thus better ventilation.

A combination of openings can direct air flow as desired.

Window units which have maximum opening areas allow good entrance of summer breezes.

Openings placed lower in the wall surfaces allow better cooling.

Exterior features can be used to direct air flow.

Window element or deflectors may be used to direct air flow.

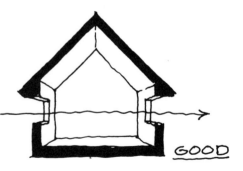

GOOD

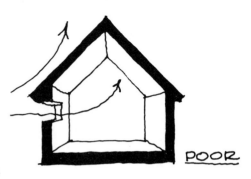

POOR

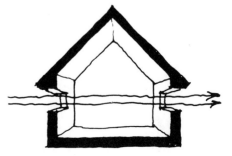

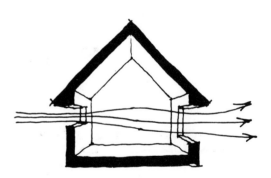

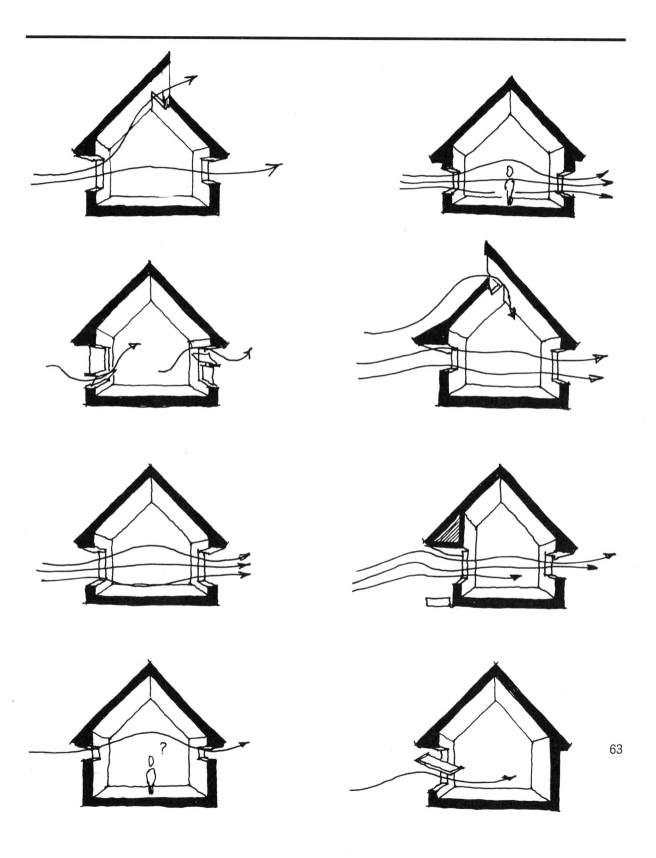

63

TIGHTENING UP YOUR OLD HOME

To keep warm out of doors there are three things to do, stay out of the wind, dress warmly and build a fire. The same advice, and more or less in that order of importance, should govern your approach to tightening up your house against cold or heat.

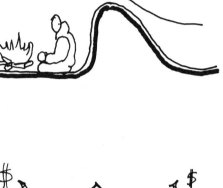

Many older homes lose a great deal of heat to the outside during cold seasons because they are inadequately insulated. Adding insulation will reduce heat loss in winter and also make the buildings more comfortable in hot weather.

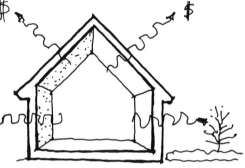

In a simple uninsulated home in good repair with normal construction but with no provision for heat conservation, about a quarter of the interior heat loss is through the roof, a quarter through the windows, a quarter through the walls, a fifth through air leakage and the remainder through the floor. Remedies vary in cost, complexity, effect. They depend as well on local conditions, state of repair and building design. Many remedies are inexpensive. Most may be performed by a careful, persistant person with limited skill and training.

(Much of the material on improving existing homes was contributed by Alfred Alk.)

Living areas should be separated from storerooms, attics, garages, crawl spaces and unheated basements. Any spaces included within the insulated envelope will share the heated inside air.

Walls may be insulated by adding blown or foamed-in-place insulation.

Basement or masonry walls may be insulated by applying rigid board insulation or by putting up furring strips and inserting blankets in the usual way. Frame walls which cannot be "blown" or foamed can be covered in the same way.

By far the greatest heat loss is through the ceiling or roof. "Capping" can reduce this heat loss significantly. This insulation may consist of blankets, batts or poured insulation placed in the ceiling framing. This will produce a cold attic which can then be ventilated.

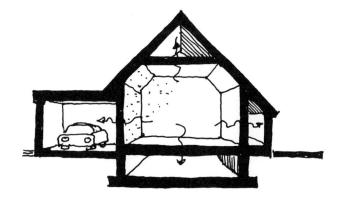

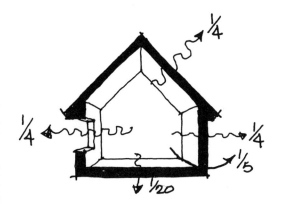

65

OPENINGS FOR NATURAL LIGHT

All exposures offer adequate amounts of natural light. The north, lacking sun, is the easiest to control and may well be the most desirable exposure for spaces with special conditions.

East and west are the most difficult to control having low-lying sun, the west being particularly troublesome.

Southern exposures, with high sun angle, offer the most light and the best opportunity to control and use it to good advantage.

Horizontal window openings are especially useful for controlling light on southern exposures.

Vertical window openings are most useful for controlling light on eastern and western exposures.

High-placed windows offer the most light distribution and the deepest penetration of natural light.

Clerestories and skylights offer good possibilities for lighting interior spaces.

Windows may be located to light specific tasks. Replace them with specific artificial light when there is insufficient daylight or at night.

Openings may be designed to light specific areas.

Smaller openings may be used to frame specific views rather than installing large "picture windows."

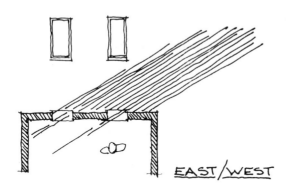

EAST/WEST

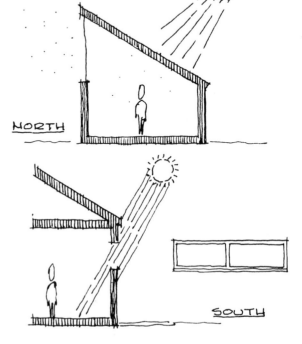

NORTH

SOUTH

66

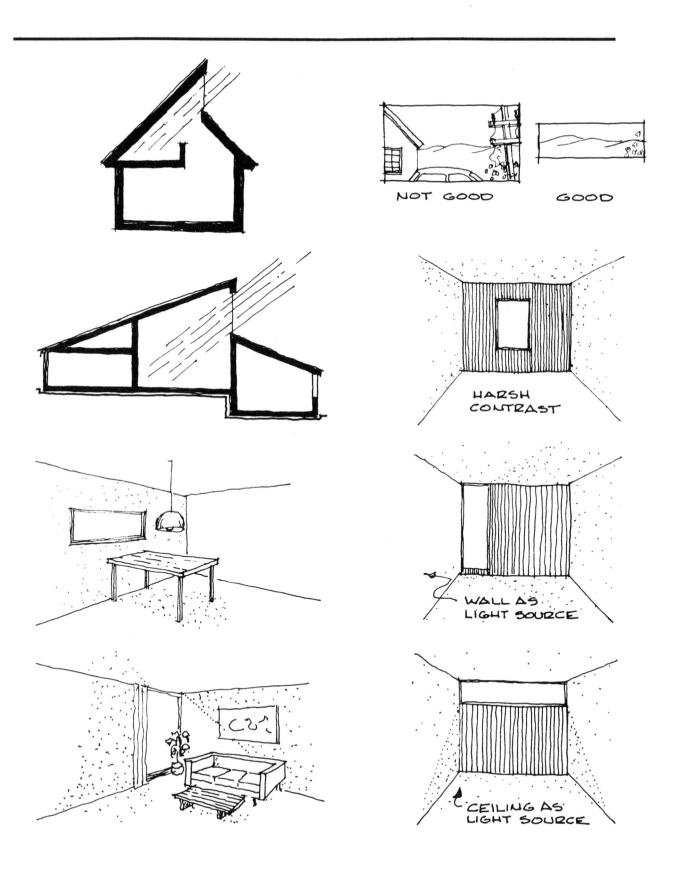

NOT GOOD   GOOD

HARSH CONTRAST

WALL AS LIGHT SOURCE

CEILING AS LIGHT SOURCE

MORE ON INSULATION

If you might wish to use the attic space, either now or in the future, batts or blankets can be installed in the roof framing. Be sure to allow for good ventilation between the insulation and the lower surface of roof. This will prevent condensation in the area and make the rooms beneath more comfortable in hot weather.

A most effective capping installation is to do both, using the whole attic area as trapped, insulating air.

How much insulation to use depends on climate, availability in local markets and house construction details. No more than 3 1/2 inches may be packed into most wall spaces. Much thicker layers may be applied over ceilings and below floors.

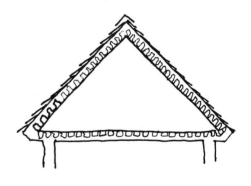

Insulate the access door to the attic and the stairway walls to the unheated basement and the attic fan louvers in the upper hallway ceiling.

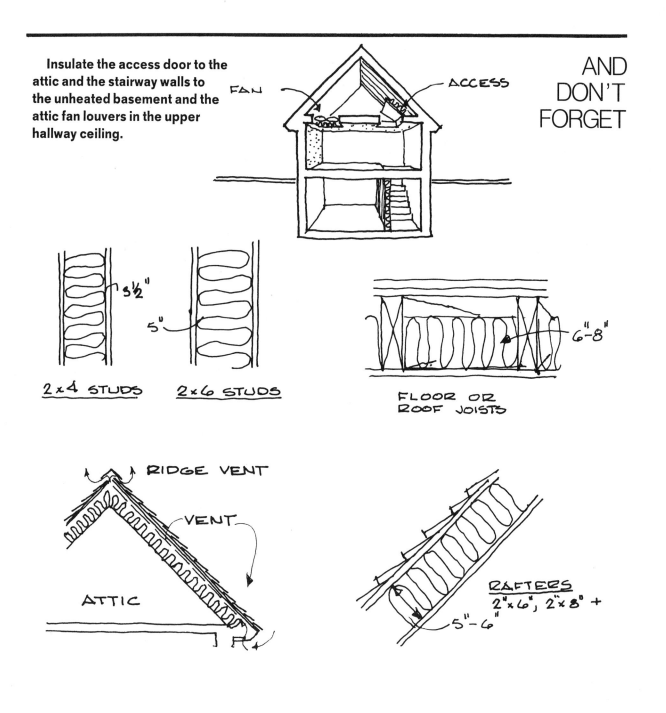

FAN

ACCESS

AND DON'T FORGET

5½"

5"

2×4 STUDS

2×6 STUDS

6"-8"

FLOOR OR ROOF JOISTS

RIDGE VENT

VENT

ATTIC

RAFTERS
2"×6", 2"×8" +

5"-6"

FIX AND CLOSE UP

Start with the obvious. Fix broken window glass. Repair and replace shingles and siding which may be missing or broken. Fix locks, latches and door closers. Refit ill-adjusted doors and windows. Cover the air conditioner; seal the opening between the upper and lower sash and around the sleeve or housing. Pile earth, bales of hay or bags of leaves against a drafty foundation, seal it with building paper or build a masonry wall under the house.

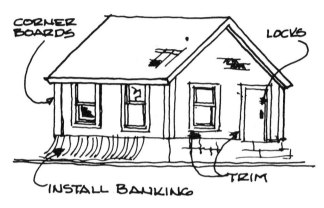

CORNER BOARDS

LOCKS

INSTALL BANKING

TRIM

WEATHERSTRIP

Weatherstrip those window openings which will be used during the cold weather. Seal unused openings with plastic roping which may be

PLAIN   W/METAL   W/WOOD

FELT WEATHER STRIPPING

BUFFER
(CLOTH OR PLASTIC)

SPONGE

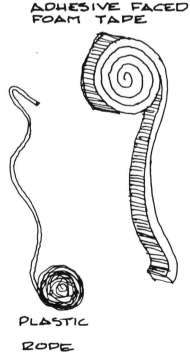

ADHESIVE FACED FOAM TAPE

PLASTIC ROPE

With a caulking gun and appropriate tools to remove hardened, ineffective caulking and dirt, look for and seal cracks on the outside of the house. These will be found around window and door frames and where siding meets the corner boards. This is a good time to make note of repairs needed to the flashing (usually metal) which covers the horizontal joint at the top of the window or door frame or around the chimney and piping. Go systematically from one opening to another.

Close up cracks and holes in the foundation. Missing mortar, old pipe channels, animal activity and water scour all contribute to problems in this area.

Next, look for air leaks in the forgotten places of the interior walls: electric and plumbing inlets, switch plates, walls plugs, ceiling light fixtures, holes between floor and basement or ceiling and attic, and in open corners or behind crown moldings and baseboard where plaster has split and never been mended. Finally stuff glass fibers or other porous material into the stud space openings in the attic and on top of the foundation wall to impede the chimney effect of cellar to roof openings.

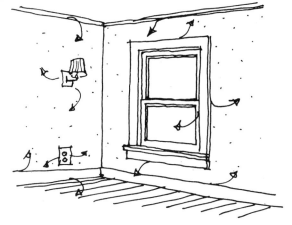

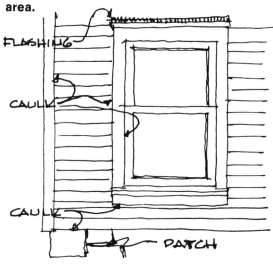

FLASHING

CAULK

CAULK

PATCH

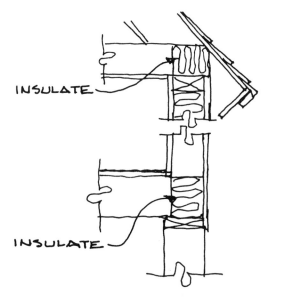

INSULATE

INSULATE

FIREPLACE

The fireplace and chimney are large heat gobblers, particularly when they are not in use. The damper should be closed whenever there is no fire. Lacking a damper, make a wooden or metal shield to fit into the flue opening, or to fit the fireplace front. If the fire doesn't burn all night, make a snuff box so the damper may be closed when you go to bed, or place a tight-fitting asbestos board plate tight over the fireplace opening.

PANEL

PANEL

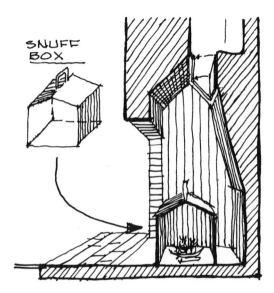

SNUFF BOX

Your house should now be tight. Stray breezes are eliminated and ventilation is under your control. Heat will still be lost against cold inside surfaces. Storm windows cut down radiative and conductive losses by impeding infiltration and by providing an air space between inner and outer glazing. Wooden sash should be loose enough around the edges for easy removal, but should be held tightly against the frame of the window. Caulking is needed under permanently-installed metal storm window frames. Two weep holes must be kept open at the bottom edges to allow for drainage and minimum ventilation, or the sills may rot.

In place of glass, transparent acrylic plastic may be used for large openings or for inside installations. Casement windows or others with metal frames must be fitted with inside storms made specially to fit the openings.

Storm doors must be well fitted if they are to be effective. Glassed decorative panels which surround front entrances frequently are overlooked when multiple glazing is being considered.

Polyethylene sheeting may be used as a covering for large openings inside and outside, but its life is usually limited to a single season.

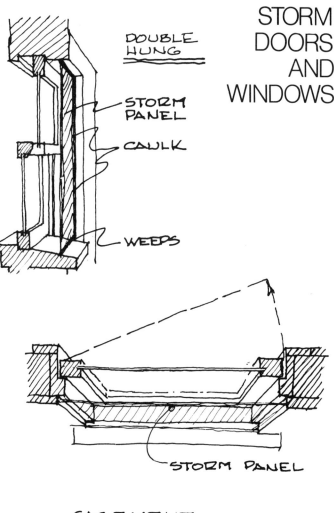

DOUBLE HUNG

STORM PANEL

CAULK

WEEPS

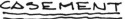

STORM PANEL

CASEMENT

FORGOTTEN PANELS

VAPOR
BARRIER

Warm air holds more moisture than cold. Inside air passing through insulation will leave moisture inside it. Where possible, therefore, a vapor barrier is placed next to the warmest or in- side of the envelope. Only one barrier should be applied. Layers of insulation added to existing insulation in ceilings and floors should be non-barrier type, or should have the barrier stripped off or slashed open. Vapor barriers cannot easily be put into existing walls, but the layers of paint on inside wall surfaces serve instead. Wall cavities may be filled by blowing or pouring in fibrous or particulate insulation, or pumping in plastic foams. Non-professional application of these latter is difficult because special equipment may be needed, particularly for the most efficient materials.

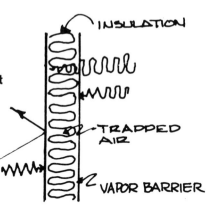

INSULATION

TRAPPED AIR

VAPOR BARRIER

If ceilings and floors are insulated, the spaces on the cold side must be ventilated. Especially in cold weather, controlled air movement must be encouraged to remove condensed moisture. In summer large air movement is needed, particularly in the attic, to keep it cool. If there is no ventilation, attic heat will be transmitted into the insulation, which will radiate the heat downward into the rooms below long after the sun has set.

VENTILATION

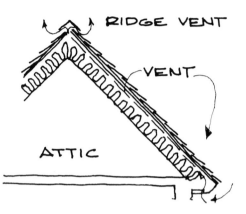

Not all the house services are included within the insulated envelope. For example, heating and cooling ducts and cold air returns and steam, water and waste piping which cross unheated attic or cellar spaces should be protected with thick insulation to protect their contents and to keep the metals from robbing heat from adjacent spaces.

RADIATION LOSSES

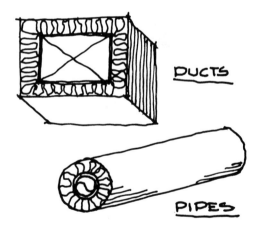

6

BUILDING SYSTEMS

Having talked about the skin and bones and the flesh of your home, let's turn now to the circulation and nervous system — in short all of the things that make it habitable and pleasant. These areas have to do with light, heat, plumbing, wiring, etc. They are the house's systems which traditionally have been the major users of energy.

It must be admitted that all of your efforts to create an energy-efficient house to this point have not resulted in a zero energy house. Regrettably, you probably will need to obtain most likely through purchase, a certain amount of energy to operate the house.

There are, however, significant choices available in buying energy, and we will suggest ways that the needs and impacts can be minimized. We also will show how your choices may relate to future opportunities, how systems can be planned to fit future sources of energy or patterns of energy consumption.

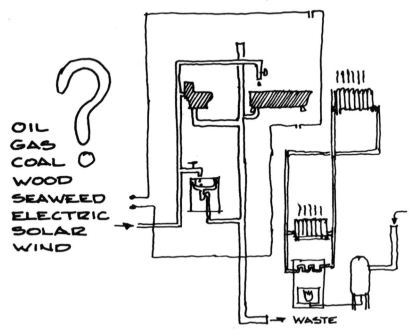

OIL
GAS
COAL
WOOD
SEAWEED
ELECTRIC
SOLAR
WIND

WASTE

OIL (including all petroleum-based distillate residual fuels), is the most common firing fuel presently in use. Its long-term availability is in question and, if available, the cost will most certainly be high. Oil's most desirable characteristic is that it is simple to transport and store.

With a thermal efficiency of 40 to 60 percent, fuel oil is considered a "high heat" firing medium and requires a substantially-constructed fire box and flue. Its housing also requires some safety and fire-resistive consideration, all of which combine to make the initial cost relatively high.

Maximum efficiency and safe operation depend upon continuing maintenance. If improperly burned, oil fuels give off unpleasant pollutants of sulfur-based compounds and particulate matter.

With its obvious drawbacks, it seems strange that oil heating remains so popular.

GAS, including utility gas and containerized liquid propane and liquid natural gas, has a source identical to petroleum, and it suffers from the same reliability and cost problems. The transmission, transportation and storage techniques are more complex, which could substantially effect the future costs and availability of gas.

Gas, with its "good" thermal efficiency of 50 to 70 percent, has a low heat-firing characteristic and requires relatively simple fire boxes and vents. Safety precautions for storage and use are stringent for obvious reasons. Gas products burn cleanly, producing few noxious by-products.

Gas offers simplicity and flexibility in operation. The basic hardware would probably be adaptable to a hydrogen or methane-based energy system if it were developed in the future.

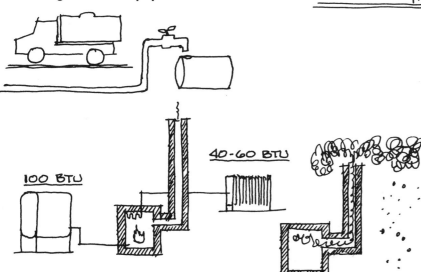

50-70 BTU

100 BTU

100 BTU

40-60 BTU

COAL is having a resurgence, particularly in North America which has most of the world's known reserves.

Coal fell into disfavor due to bulkiness in handling and unclean firing characteristics. It is a high heat medium, requires substantial fire box and flue construction, and a controlled supply of combustion air.

If improperly fired, coal gives off quantities of lethal carbon monoxide, so a carefully-designed system with safety devices is mandatory. In combustion coal also gives off carbon and sulfur-based pollutants, particulate matter (soot) and leaves an ash residue which is messy to dispose of.

While it offers greatest availability and most attractive long-term cost, coal's best use is for large scale operations, for heating and producing electricity. Domestic use, while completely feasible, should be carefully evaluated.

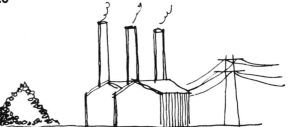

ELECTRICITY is considered not so much an energy source as a transmission medium, since its source usually is hydroelectric (water powered), nuclear or fossil fueled (coal, gas, petroleum) generating stations.

Current electric costs are modest (depending somewhat on region and location). However, future cost and availability will depend on construction of generating and transmission facilities and sources of fuel. All fuel sources, whether fossil, nuclear or exotic, are subject to serious drawbacks at present.

Electricity's overall thermal efficiency is low—25 to 35 percent, but equipment and installation are extremely flexible and relatively simple. The energy delivery is restricted to resistance heating coils which have a high surface temperature requiring protection and care in use.

An electrical system which could be hooked to a reliable or ample local generating source would appear to be a most logical long-term approach. Lacking the re-liability of source, electricity as a complete energy system appears questionable and ultimately expensive.

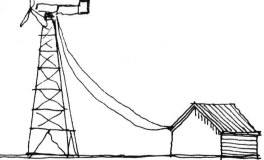

ON-SITE FUELS

The fuels which may be available on or near your site generally are labor intensive and depend on reducing organic matter to a useable form and burning it for its heat content. The oldest and simplest is wood.

Carefully-selected species of wood have a high thermal efficiency, burn cleanly, leave a little ash (which is good fertilizer) and produce no known pollutants aside from minor particulate matter.

Wood declined in popularity and use due to the hard work in "getting it out", and problems in storage and transportation. The former has been partly overcome with development of small, mechanized harvest equipment.

Another major drawback to wood has been the lack of adequate technology for burning. The most familiar method is in an open fireplace, which is notably inefficient. Even in modern stoves and furnaces the lack of uniformity of the fuel imposes serious efficiency problems. Fireplaces should be considered primarily for their aesthetic appeal and only secondarily or in an emergency as a heat-producing source.

There are numerous stoves and a few furnaces with a thermal efficiency approaching 50 percent which make good use of wood, and more are on the way.

The incomplete burning of wood causes deposits of combustible materials in flues and stacks which will burn unexpectedly (chimney fires) if not cleaned out regularly.

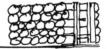

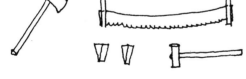

90 BTU

10 BTU

50-60 BTU

100 BTU

This must be considerd a major hazard to be dealt with accordingly. Pipes, stacks and flues should be carefully designed, built and maintained. The open flames, embers and hot surfaces of stoves can be a fire and personal hazard and must be carefully considered in any proposed use or design.

The burning of other organic matter produced on the site is similar to wood and is a matter of individual circumstance. The burning of peat, dried corn stalk, pressed straw, sea weed and dried animal waste are possible, and if available in sufficient quantity, certainly should be considered.

The conversion of wastes such as garbage and manure requires large quantities and complex processing in order to yield useable amounts of fuel gas. These should be considered if volume of waste and the processing is available. See Chapter 8 for more on this.

CREOSOTE

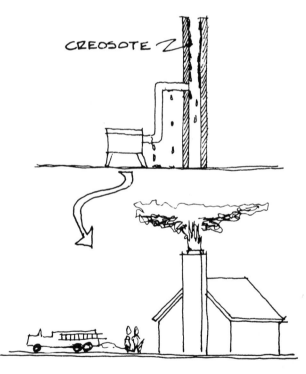

HEATING METHODS

Here are the common methods for moving heat around within your home. Note also how they might adapt to future technologies.

FORCED HOT AIR is the simplest and most direct, requires electricity to keep the air moving. The system allows for filtering and humidification. Since it sometimes depends on large and expensive duct work to circulate the air, it is most adaptable to simple, compact plans. It is adaptable to solar heating systems and wood-burning furnaces.

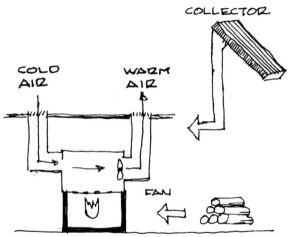

CONVECTION HOT AIR

Here the heated air is circulated by natural convection currents and does not depend on electricity. In simple applications it is surprisingly effective, and promises to become as popular as it once was. Convection hot air does require careful design, both in building plan, spaces and circulation system. See Chapter 8. for more details.

FORCED HOT WATER

Here heated water is pumped (by electricity) to radiation elements. It is clean, quiet and very effective. Humidification if desired, must be added as a separate system. A forced hot water layout is adaptable to many of the current solar heating systems.

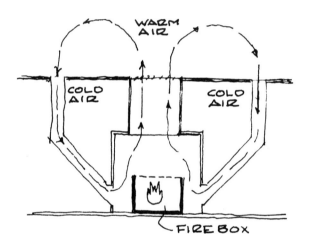

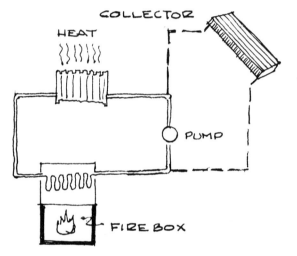

ELECTRIC RESISTANCE

This consists of coils wired to an electrical source. It is clean, simple, flexible, but sometimes the heating element surfaces get very hot. It provides very dry heat, and humidification is usually necessary. As a heating system it is adaptable to wind or other locally powered electrical sources. Such modified source units must be of large size.

HEAT PUMP
In areas where winter weather is mild, this system may be used where cool outdoor air chills refrigeration coils. The liquid is compressed, giving off heat to the house. In summer the unit can be reversed to cool the house. Electricity is the fuel used to create mechanical energy which then converts unuseable heat to useable heat.

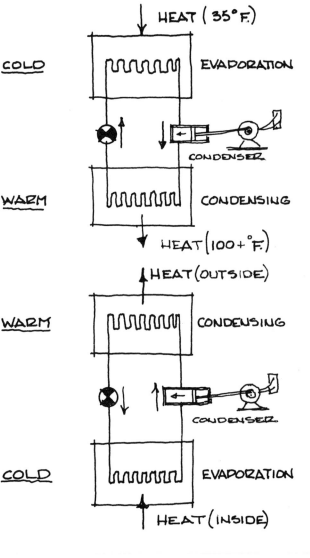

HEAT (35° F.)

COLD — EVADORATION

CONDENSER

WARM — CONDENSING

HEAT (100 + °F.)

HEAT (OUTSIDE)

WARM — CONDENSING

CONDENSER

COLD — EVADORATION

HEAT (INSIDE)

Use natural ventilation insofar as possible. Where forced ventilation is necessary or highly desirable use as little as possible. Consider ductless range hood for kitchen, for instance, to return heat of cooking to heating requirement of the home.

If you're considering air conditioning and unless you have a health need in the family, you obviously haven't studied the alternatives suggested in earlier chapters sufficiently.

For homes located in extremely hot and dry climates consider water evaporation air cooling. Hot but humid situations may be greatly benefited by dehumidifiers alone.

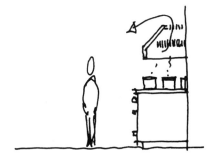

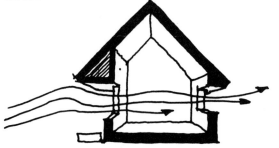

LIGHTING & APPLIANCES

We should acknowledge that a self-sufficient way of life probably is possible only by using a wide variety of power-operated labor saving appliances.

Power tools allow you to build or improve your home quickly and well. Household appliances assist in preparing food and providing comfort which you doubtless need and deserve. Still others perform tasks which will free you to pursue other areas of self-sufficiency. It's a matter of "trade offs."

But beyond this, the line has to be drawn. Many appliances are sheer waste, an excuse to manufacture and sell a product and create a demand for energy forever. Learn to know the difference.

In the area of light and illumination the watchword is "don't." Getting about outdoors after dark can be accomplished at light levels of $1/2$ to 5 footcandles. Indoors seems to need 5-15 footcandles. Most tasks require 50-75 footcandles, rarely more than 100.

In spite of this many public buildings, public spaces and individual homes are lighted to several times these levels, obviously an unnecessary waste.

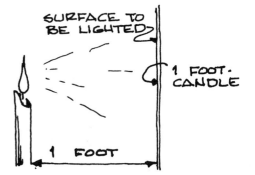

SURFACE TO BE LIGHTED

1 FOOT-CANDLE

1 FOOT

50-75 F.C.

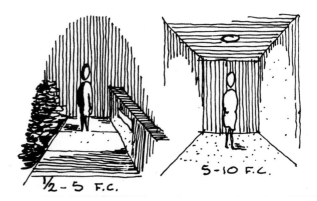

$1/2$ - 5 F.C.

5-10 F.C.

86

There are two types of light bulbs generally available for residential use.

Fluorescents are very efficient in terms of light per unit of energy used. But the quality of the light tends to be unpleasant to the eye, and the form (tubes in various lengths and shapes) is not flexible. Attempts to conceal the light source and cut down harshness have the effect of wasting light and thus energy.

Incandescents are less efficient with added energy used going to produce heat. But the light characteristic is warm and pleasant, and you can consider the heat generated as part of heating requirement for space.

Conclusion: Use fluorescents to light large areas or specific surfaces. Use baffles or reflectors to shield the light source from the eye and to condition the quality of light. Use incandescents to light specific tasks or for pleasant effect. Get as close to the task or surface as possible to use the smallest bulb and produce less heat.

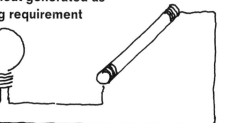

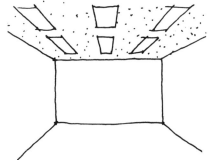

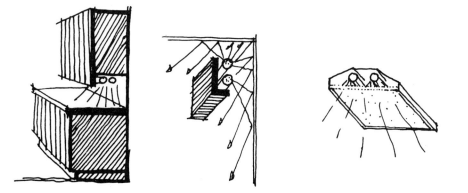

If large amounts of light are necessary, consider the heat generated as part of heating requirement.

Here are a few things to keep in mind when developing ideas for low-energy consumption lighting.

You need to see what you are doing, but only the immediate task or work area needs to be lighted to the full, required intensity.

In perceiving objects in a space, the light level in the space or on individual objects is not so important as is establishing some contrast or difference. Use backgrounding and silhouettes to emphasize three-dimensional aspect or for definition. However, extreme contrasts between light levels are very fatiguing and should be avoided.

NO CONTRAST

EXTREME CONTRAST

7

Here are some ideas that will help you make good on the observations and ideas that you developed earlier regarding your site.

They also will follow through on the decisions made in your energy-efficient house itself, allowing you to save more dollars and energy, as well as help you make greater use and derive greater enjoyment from your land.

Probably you will agree that all of the desires you have developed have little chance of being accomplished all at once. Nor should they.

Instead, you should draw up a long-term master plan with complete phase schemes both for your house and its surroundings.

Working on your land is one instance where time is on your side. There is time to work things out properly, time to plan carefully. Meanwhile, anything that you have planted presumably is growing and adding to the enjoyment of your place.

Decide now what you ultimately want to achieve. Do now what you need to in order to maintain yourself. And enjoy watching the balance grow.

89

A REVIEW OF SITE NEEDS

Let's again pick up what you developed in your site plan in Chapter 2. Our sample plan had to do with plantings for wind protection, for shading and heat tempering, for screening and privacy for all the things you wish to do on the land. It was left in the form of a diagram.

Use your plan for this as the "program" for your master plan for planting and the exterior furnishings of your energy-efficient house.

In working with your existing home, this is one area where you have a good opportunity to make your homestead an energy-saving and highly productive one. In some cases the exterior and grounds may present the greatest opportunity.

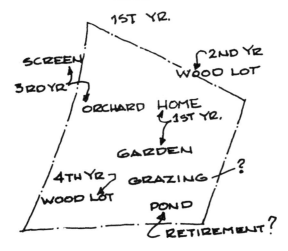

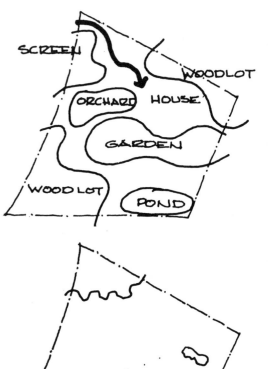

Do not overlook the possibility of fulfilling planting needs with nut and fruit trees and shrubs. Here is the opportunity for trees to provide decoration, wind protection and tempering, and also to yield food, to say nothing of the joy of working with them. Even the wood of these trees is good for small craft work and it makes fine fireplace fuel. In cooler climates the sap of sugar maples can provide the basis for a small home industry or at least provide syrup and sugar for home use.

Thus groves of trees, providing wind protection, screening or decorative quality, can also yield apples, pears and cherries and many kinds of nuts in season. Hedges erected for screening or snow control can give you berries and nuts too.

POOR   GOOD

LARGE PLANTINGS – DECIDOUS

Deciduous trees, that lose their leaves in cold weather, are excellent for shade in the summer, but are open in winter. These are good two-way trees, allowing desirable sun gain in winter and providing shade when it's needed.

Deciduous trees provide limited wind protection in winter except in large clumps or in combination with other types. They can be most effectively employed to modify local breezes in summer. Other characteristics to consider are:

Leaves provide a good source of mulch materials, harvested once a year.

Most species provide good to excellent fire wood and also can be selected for wood-working purposes.

Densely-shaded species discourages growth on the ground directly below — good places for growing shade-seeking ground covers.

Sometimes they present a hazard near buildings as branches break off in wind storms and under snow and ice loads.

Foliage turns spectacular colors in autumn in some locations.

The structure of the bare tree without leaves is of visual interest.

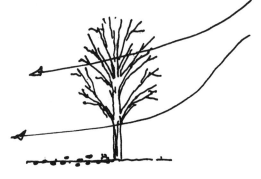

Coniferous or evergreen trees, as the name implies, are always green. As such they are good for major windbreaks and baffles and for visual and privacy screening. They are the only thing available if you want or need permanent green all year.

Large evergreens have a characteristic structure or single stem and radiating boughs. Smaller ones may have a diversity of structure. The wood has limited value for burning but is good for timbers, poles or as saw logs. The greens and cones are well known for decorative quality.

Remember that the needles of most coniferous species are acid and will affect the soil surrounding. If a ground cover is to be maintained, the old needles will have to be raked up each fall.

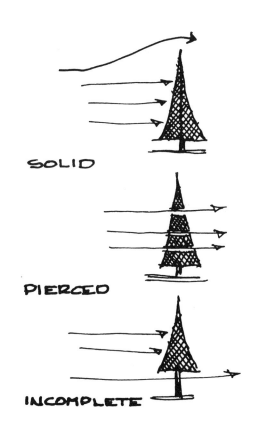

SOLID

PIERCED

INCOMPLETE

93

WIND PROTECTION – LARGE SCALE

Shelter belts are large masses of trees planted to reduce wind velocity for a substantial area, perhaps for your whole homestead or a cluster of homes or buildings. The belts should be dense plantings of evergreen trees, preferably in two or three rows.

Windbreaks can be used to protect individual areas or elements from the full force of the wind, both to the windward and leeward side of the windbreak barrier. They can control to windward for a distance of 2 to 5 times the barrier's height and to leeward a distance of 10 to 20 times.

Windbreaks and wind barriers also may be earthen structures such as berms and mounds, masonry walls, wood fences or hedge plantings.

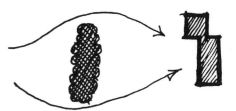

POOR

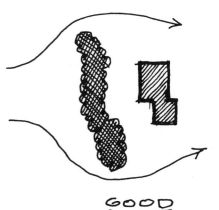

GOOD

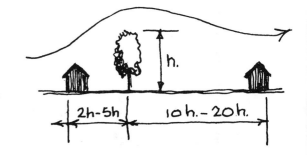

On a smaller scale, landscape elements are used to shelter buildings both by redirecting the full force of the wind or by creating dead air spaces next to building surfaces, near doors or windows, or outdoor spaces which you wish sheltered.

Wind velocity, which increases at corners of buildings, will be reduced by barriers or plantings in these locations.

On the smallest scale, vines on a wall surface will provide dead air space for winter protection and shade for the wall in summer. Moisture from the plants also will cool by evaporation. The same moisture, however, may cause deterioration of certain surfaces. Use of vines, therefore probably should be used only on durable materials such as masonry or concrete.

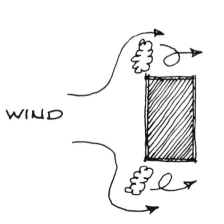

WIND PROTECTION – SMALL SCALE

DEAD AIR SPACE

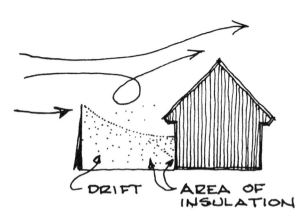

DRIFT — AREA OF INSULATION

SHADING

Large trees, particularly the large deciduous varieties, are useful for shading building surfaces and openings. Obviously, southeast to southwest orientations are best. Deciduous trees provide the most shading in summer and allow maximum solar radiation heat in winter.

Evergreens can be used for effectively shading the low-lying sun and can perform dual function of wind sheltering.

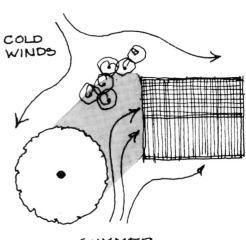

COLD WINDS

SUMMER BREEZES

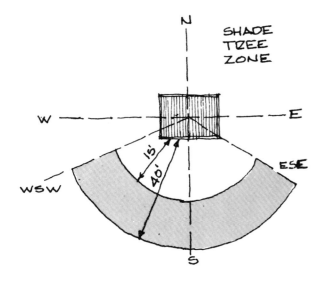

N

SHADE TREE ZONE

W

E

WSW

15'

40'

ESE

S

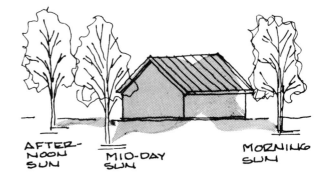

AFTER-NOON SUN

MID-DAY SUN

MORNING SUN

On a more positive note, plantings may be used to enhance natural ventilation. This is particularly effective when adapting existing structures that have fixed window openings with somewhat less than desirable orientations.
For example:

Air flow can be increased or directed through the building.

Tall hedges can effect major changes in the direction of air flow.

Combinations of large and small plantings can be used for different flow patterns.

The overall planting scheme can provide a whole new pattern of natural ventilation.

Large plantings can be used to re-direct the breezes on the site.

Smaller plantings can be combined with building openings, either new or proposed, to get a better circulation of air within buildings.

NATURAL VENTILATION WITH PLANTINGS

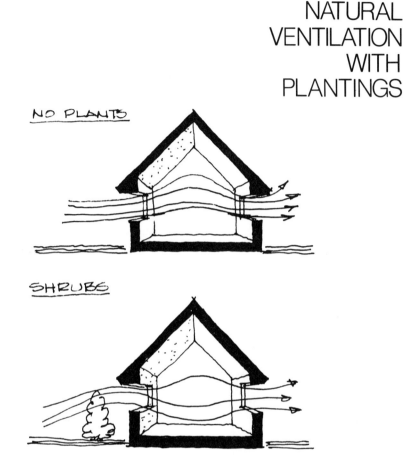

NO PLANTS

SHRUBS

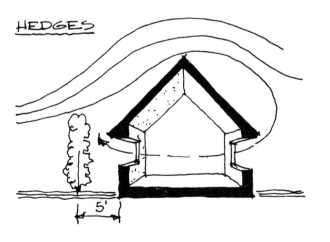

HEDGES

5'

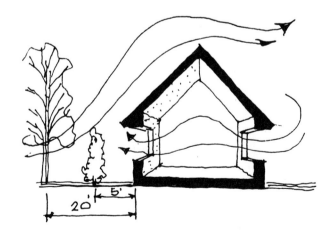

20' 5'

TEMPERING

The environment adjacent to buildings can be tempered effectively by the nature of surrounding surfaces. This also will prevent reflected heat from entering the house. Ground cover and low shrubs are most effective and grasses are next. Gravel and exposed soil are relatively poor. Sheet asphalt is the worst, causing a large amount of re-radiation and absorption.

In the winter surfaces covered with snow can be an additional source of reflected solar light and heat.

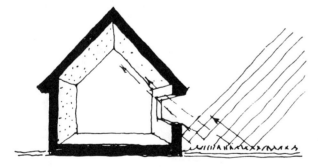
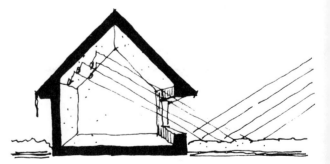

Taken all together your finished plan may look something like this:

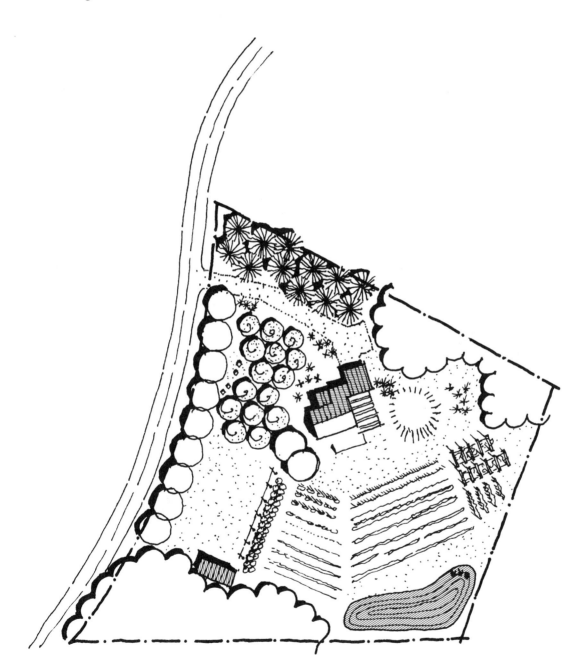

8

NEW DIRECTIONS

Our final step will be to familiarize you with new or upcoming ideas to conserve energy and money in home building. It will not so much provide full information but rather hints of what to look out for in the future.

These new, old, or emerging technologies realize further conservation of existing energy modes and ways in which one can make greater use of natural, free energy sources. Perhaps most interesting are the ideas for recycling or saving energy in ways which always have been with us, but perhaps have been almost forgotten.

WASTE

The easiest and simplest way to use solar energy is to let the sun shine on the area to be heated — on your roof, in your living room window or on you. This, however, may not be the most efficient. It may be somewhat ineffective because the sun shines only a part of a 24 hour day and not every day. It also is inefficient because, as observed earlier, surfaces should be specifically designed in order to receive maximum solar radiation.

First, to heat by solar radiation, you need a collector, an assembly especially designed to absorb the sun's heat.

Next you need a way in which to transfer the heat from the collector to the place where you will use it. This is usually done with a fluid medium, water, or some other fluid, or air.

To take care of heating needs when the sun isn't shining, there should be some way to store the heat.

The manner in which the solar heat ultimately benefits you could be any of the heat-transfer methods discussed in Chapter 6.

Other equipment which may make the system more useable is a "heat pump" or other supplemental heat sources to see you through the periods when the sun cannot provide all of your requirements.

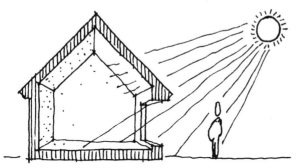

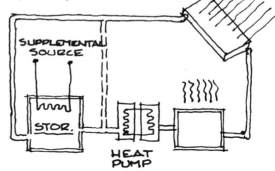

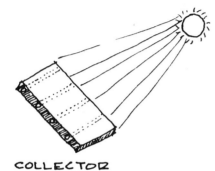

COLLECTOR

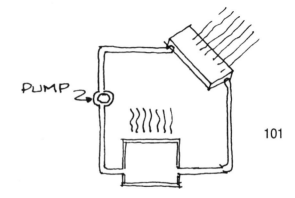

101

WIND POWER

Another popular energy-producing source is the wind. Wind power derived from mills has been used for centuries, providing mechanical energy for useful work, notable to pump water. More recently the mechanical energy has been converted by generator to electrical energy.

The shortcoming here is what to do when the wind is not blowing. The common answer is storage batteries, which at present are cumbersome, expensive and relatively short-lived.

On a domestic scale another immediate drawback is that the quantities of power needed, at the voltages normally required, are not available from simple systems. To operate a conventional "all electric" house would require a mind-boggling number of wind mills and a huge array of batteries.

But by confining electric use to those areas most necessary, or which offer the most convenience, a workable wind power electric system can be set up.

A suggested large-scale solution to the storage problem would use wind-generated electricity to convert ordinary sea water into its component parts, hydrogen and oxygen, by electrolysis. The hydrogen could then be piped, stored and used as an ordinary gas fuel. More efficiently, the hydrogen and oxygen could be recombined in a fuel cell for ultimate use to produce water and electricity (the reverse of electrolysis).

The practicality and reliability of fuel cells has been demonstrated on space flights where they have provided a major portion of the electricity. Before the process is feasible for widespread use on earth many problems of cost and operation must be solved.

Another way: somebody blessed with a strong wind source, ample water supply and a large drop in elevation could consider a wind-powered pump storage system.

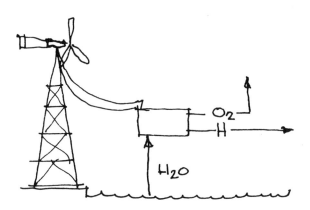

COMBINATION PUMP / TURBINE & MOTOR / GENERATOR

O_2
H
H_2O

ELECTRICITY

Electricity is used also to operate a "heat pump," referred to earlier, which basically is the unit found in an ordinary refrigerator, but with a difference. The heat pump also is useful for recovering low-grade heat or excess heat and recycling it to better use.

In these cases electricity is used to produce mechanical energy, which in turn extracts heat energy at a higher rate.

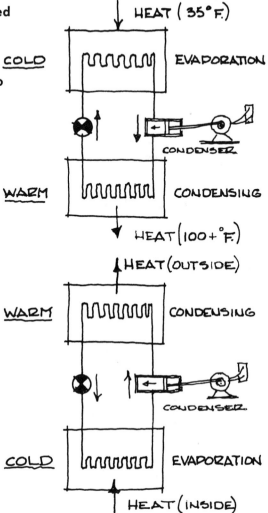

HEAT (35°F.)

COLD   EVAPORATION

CONDENSER

WARM   CONDENSING

HEAT (100+°F.)

HEAT (OUTSIDE)

WARM   CONDENSING

CONDENSER

COLD   EVAPORATION

HEAT (INSIDE)

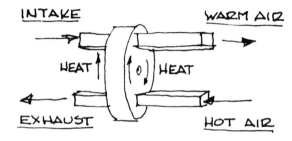

INTAKE   WARM AIR

HEAT HEAT

EXHAUST   HOT AIR

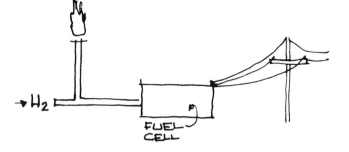

H_2

FUEL CELL

New programming by the electric industry might allow cash savings for purchased energy. It would stem from the fact that at certain hours of the day electricity is being generated for which there is no user demand — early in the morning when most everyone is asleep.

Some utilities offer this electricity at rates substantially below normal, and there is every indication this practice will increase. The task is to buy the energy when it is least expensive and store it for the time when you can make best use of it. This is done most simply by heating water (or other fluid) in a large tank. The tank holds the heat with little thermal loss until you need it, and then is drawn off.

There are other means of recovering heat for reuse (or cooling) by using mechanical energy in heat exchangers. The ready example is a "heat wheel" which recovers the heat in air which is to be exhausted into the outside and otherwise wasted. The same equipment can be used on the other end of the cycle and recycle cool air back into a building.

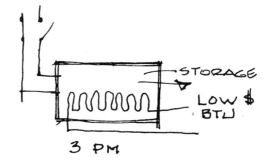

3 PM

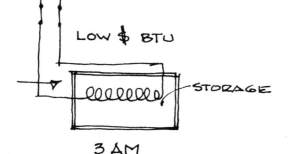

3 AM

ORGANIC SOURCES

Other simple sources of energy include garbage, human or animal wastes which are converted to useable methane gas by a variety of methods, most notably a digester.

The extraction of methane does not detract from the value of the basic material for fertilizer—truly an efficient use of something which might otherwise be considered waste. It should be emphasized, however, that relatively enormous quantities of waste are required to make this an economical system and to provide a significant amount of energy.

A simpler more direct use of the same material is through composting.

While this does not yield large amounts of energy in the conventional sense of the derived gas, it does produce ample quantities of fine fertilizer which can ease the demand for fertilizers derived from expensive and limited petroleum supplies.

The heat generated by composting can be of use. While we do not suggest heating your home from your compost pile, it might be used for starting plants or in "fueling" a hot house or "hot frame."

METHANE

COMPOST

A relatively new approach to energy conservation is what can be called "super insulation." The basic principal is retention of heat in an enclosed volume, such as the way an ordinary refrigerator or freezer holds cold. It depends on very efficient insulation or sealing of the volume to be heated (or cooled).

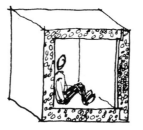

Using an example of a 50 degree differential in temperature, normal construction would lose about 10 BTU/hr. per square foot of building surface. Applying the care and standards found earlier in this book would reduce this to about 4 BTU/hr. Using the principals of super insulation would reduce this to $1\frac{1}{2}$ BTU/hr. — a truly remarkable saving.

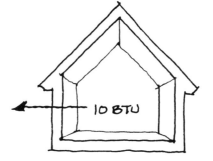

Super insulation depends on cellular foamed insulations usually put together in "sandwich" units which then are carefully joined together to form an extremely tight and well-insulated enclosure for spaces.

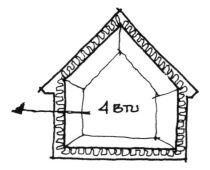

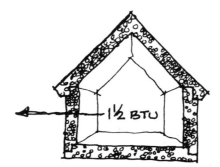

GRAVITY AIR DISTRIBUTION

The "gravity" heating system is back! The advantage is that you don't have to use extra energy to distribute the heat around the house, letting natural convection currents do the job. It does require that the whole house, particularly the heating system, be carefully designed to provide continuous and comfortable circulation of air.

Cold air at the walls drops through vents in the floor, is moderated by warmer earth temperatures and then rises past the heat source to continue the cycle. An added benefit can be derived by using the earth adjacent to the basement for storage of heat, taking advantage of its thermal capacity. Ventilation for summer months is taken care of by natural means described earlier.

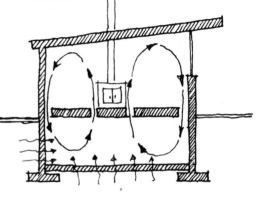

SKY THERM COOLING

Excess heat may be dissipated in an energy-efficient way through radiation to the night sky — the so called "sky therm" effect. The excess heat is stored in a fluid during the warm, daylight period and is radiated to the dark night sky. When the daylight cycle arrives the fluid is cooled off ready to receive another day's worth of unwanted heat.

HEAT TO NIGHT SKY

The characteristic of massive building materials to retain heat over long periods of time is called thermal capacity. These materials are commonly concrete and masonry but may be extended to include steel structure, the plaster interior partitions, even the metal equipment and furnishings of a building. It may be done deliberately by placing a large bed of sand under a building, sand being an excellent retainer of heat. The earth adjacent to a building may be used, also.

An old example of this application is the traditional Cape Cod house with the massive central chimney structure. The heat source was the fires in the several fireplaces. The chimney and fireplace masonry retained the heat to be distributed evenly long after the fires were banked.

The same principle is used in the massive masonry or adobe walls of traditional structures in high arid climates where days are hot and nights cold. The walls absorb heat and shade the interior spaces during the hot daylight hours, storing the heat then to re-radiate it to the inside during the cool nights — an effective use of both ends of the cycle.

Living patterns must be adjusted to take into account this principle, but sitting outside in the evening can be a most pleasant part of the day.

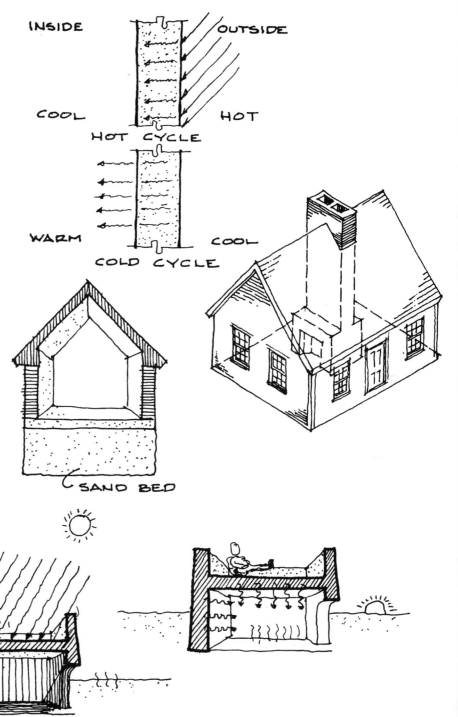

INSIDE OUTSIDE

COOL HOT

HOT CYCLE

WARM COOL

COLD CYCLE

SAND BED

EPILOGUE

There will come a time when we have learned to reduce our requirement for energy by carefully fitting our shelter to our actual needs, when we plan and design for climate, for individual building sites, using the most effective materials and readily available energy sources.

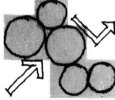

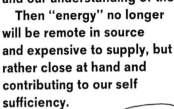

Then the size, shape and appearance of our shelter will be the result of natural forces and our understanding of them.

Then "energy" no longer will be remote in source and expensive to supply, but rather close at hand and contributing to our self sufficiency.

SELECTION OF MATERIALS

The choice of materials to build your energy-efficient house should be based on effectiveness for enclosure from weather, degree of insulation offered, durability of surface (both inside and out), local availability, ease of construction (if you choose to do much of the work yourself), attractiveness and, lastly, cost.

Unfortunately there is no known single material which does everything well, although some do a number of things adequately. Some do a single thing very well.

Generally a compromise is in order to fill a number of needs with a single material. More commonly a combination of different materials best answers all needs. Following are tables that will help guide you in selecting building materials and wall assemblies.

These tables are included to provide a general guide to the quality of common materials and their suitability for various uses. It also indicates a degree of skill or technical knowledge required for installation as well as a scale of costs.

Actual selection of materials should depend on intended use, local availabilty & cost, and necessary skills for installation.

Regional or local variations on materials and skills should be carefully considered in final selection.

A few notes:

Sheds Water—refers to shedding of water on sloped or horizontal surfaces as appropriate.

Resists Water—refers to resisting of water or moisture on vertical surfaces.

Fire Resistivity—a general combination of factors considering whether a material supports combustion, promotes spread of flame or produces harmful smoke.

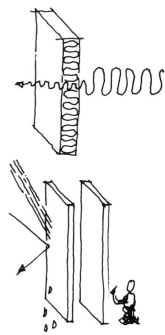

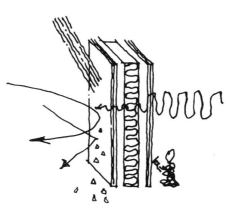

	STRUCTURE		SHELTER		ENERGY SAVING				FINISH QUALITY				APPEARANCE		COST	DIFFICULTY	REMARKS
	beams, joists & rafters	walls & columns	sheathing/decking	sheds water	resists wind	resists water	resists infiltration	thermal capacity / insulation value	fire resistivity	durability exterior	durability interior	maintenance	exterior	interior	cost	difficulty	
STONE																	
field stone	G	—	—	U	G	G	P	E	E	E	E	E	E	E	D	H	
dressed stone	E	—	—	U	G	E	P	E	E	E	E	E	E	E	D	XH	•
cut stone	E	—	—	P	G	E	P	E	E	E	E	E	E	E	D	XH	
slate-floor	—	—	—	A	A	G	P	E	E	E	E	G	E	E	LD	HM	
slate-roof	—	—	—	E	A	A	P	G	E	E	E	E	E	E	D	H	
BRICK																	
walls	E	—	—	E	E	E	P	E	E	E	E	E	E	E	LD	HM	
paving/floors	—	—	—	A	A	—	P	E	E	E	E	G	E	E	LD	HM	
CONCRETE																	
reinforced	E	E	—	A	E	E	P	E	E	PE	PE	G	PE	PE	LD	M	
masonry units (blocks)	G	—	—	U	A	E	PA	GE	E	P	A	A	P	A	LD	M	
precast	E	E	E	E	E	G	P	E	E	E	E	E	GE	GE	D	MH	
TILE																	
structural	G	—	—	A	G	E	A	GE	E	E	G	E	A	A	LD	MH	
roof	—	—	—	E	G	G	P	G	E	E	—	E	E	E	D	H	
wall surfaces	—	—	—	G	E	E	P	—	E	E	E	E	E	E	LD	MH	
floor/paving	—	—	—	G	—	—	P	A	E	E	E	E	E	E	LD	M	
EARTH																	
rammed	G	—	—	A	G	E	G	E	E	PA	P	P	AG	UP	S	LM	
adobe	G	—	—	A	G	E	G	E	E	PA	P	P	AG	UP	S	LM	

• Look for recycled.

	REMARKS	COST	DIFFICULTY	APPEARANCE interior	APPEARANCE exterior	FINISH QUALITY maintenance	FINISH QUALITY durability interior	FINISH QUALITY durability exterior	FINISH QUALITY fire resistivity	ENERGY SAVING thermal capacity	ENERGY SAVING insulation value	SHELTER resists wind infiltration	SHELTER resists water	SHELTER sheds water	STRUCTURE sheathing/decking	STRUCTURE beams, joists & rafters	STRUCTURE walls & columns
STEEL																	
rolled sections		H	D	G	G	P	G	A	P	PA	—	—	—	—	—	E	E
light framing		MH	D	P	P	UP	P	P	U	P	—	—	—	—	—	E	E
corrugated sheet		M	LD	A	A	P	P	P	U	P	U	G	E	E	E	—	—
galvanized sheet		MH	LD	A	A	G	E	E	U	P	U	G	E	E	E	—	—
painted sheet		M	LD	A	A	G	G	A	U	P	U	G	E	E	E	—	—
COPPER																	
roof		XH	D	E	E	E	E	E	G	—	—	E	E	E	—	—	—
flashing		XH	LD	E	E	E	E	E	—	—	—	E	E	E	—	—	—
ALUMINUM																	
roof		LM	S	A	A	G	—	E	A	—	—	G	E	E	—	—	—
flashing		L	S	A	A	A	—	G	U	—	—	G	E	E	—	—	—
painted siding		MH	S	—	PA	G	—	*	P	P	AG	E	E	A	—	—	—
GLASS																	
walls, windows	●●	M	LD	E	E	E	E	E	PA	U	U	E	E	E	—	—	—
skylights	●●	M	LD	E	E	E	E	E	PA	U	U	E	E	E	—	—	—

* Unknown

●●Joints & sealing are weak point.

KEY

A	Average or Adequate	**M**	Medium (cost)
D	Difficult (installation)	**MH**	Medium High (cost)
E	Excellent	**ML**	Medium Low (cost)
G	Good	**P**	Poor
H	High (cost)	**S**	Relatively Simple (installation)
L	Low (cost)	**U**	Unsatisfactory
LD	Less Difficult (installation)		

Quality Characteristics

A Average or Adequate
E Excellent
G Good
P Poor
U Unsatisfactory
— Not used or not applicable
A/G Combination indicates general range, i.e. Average to Good.

Difficulty of Installation

D Difficult, Requires special skills
LD Less Difficult, requires certain skills which may be developed.
S Relatively Simple, requires ordinary level of skill or ability.

Cost

X H Extra High
H High
M/H Medium High
M Medium
L Low

113

INSULATION / PLASTER / ASPHALT

Material	STRUCTURE: beams, joists & rafters	STRUCTURE: walls & columns	SHELTER: sheathing/decking	SHELTER: sheds water	SHELTER: resists water	ENERGY SAVING: resists wind infiltration	ENERGY SAVING: insulation value	ENERGY SAVING: thermal capacity	FINISH QUALITY: fire resistivity	FINISH QUALITY: durability exterior	FINISH QUALITY: durability interior	APPEARANCE: maintenance	APPEARANCE: exterior	APPEARANCE: interior	DIFFICULTY	COST	REMARKS
INSULATION																	
foamed plastics	—	—	—	—	A	A	A	E	P	UE	U	U	—	—	S	L	•
expanded glass fiber (fiberglass)	—	—	—	—	—	A	E	P	E	U	U	—	—	—	S	LM	
expanded mineral fiber (rock wool)	—	—	—	—	—	A	E	P	E	U	U	—	—	—	S	LM	
expanded mineral (vermiculite)	—	—	—	—	—	A	E	P	E	—	—	—	—	—	S	LM	
organic fiber	—	—	—	—	—	A	E	P	E	—	—	—	—	—	S	LM	
paper	—	—	—	—	—	E	P	—	PA	—	—	—	—	—	S	L	
metal foil	—	—	—	—	—	E	G	—	A	—	—	—	—	—	S	L	
PLASTER																	
exterior (Stucco)	—	—	—	P	G	E	PA	G	E	E	—	E	E	—	D	MH	
interior	—	—	—	—	—	E	A	G	E	—	E	E	—	E	LD	H	
board (gypsum board)	—	—	—	—	—	E	A	A	GE	—	G	G	—	G	S	L	
ASPHALT																	
shingles	—	—	—	E	E	E	PA	—	AE	G	—	E	A	—	S	L	•
membrane/"built up roof"	—	—	—	E	E	E	P	—	AE	G	—	G	A	—	D	M	
membrane (roll roofing)	—	—	—	G	G	E	—	—	A	A	—	A	A	—	S	L	•

Verify fire characteristics •

KEY

Medium (cost)	**M**	Average or Adequate	**A**
Medium High (cost)	**MH**	Difficult (installation)	**D**
Medium Low (cost)	**ML**	Excellent	**E**
Poor	**P**	Good	**G**
Relatively Simple (installation)	**S**	High (cost)	**H**
Unsatisfactory	**U**	Low (cost)	**L**
		Less Difficult (installation)	**LD**

Cost		Difficulty of Installation		Quality Characteristics	
Extra High	**XH**	Difficult, Requires special skills	**D**	Average or Adequate	**A**
High	**H**			Excellent	**E**
Medium High	**M/H**	Less Difficult, requires certain skills which may be developed.	**LD**	Good	**G**
Medium	**M**			Poor	**P**
Low	**L**	Relatively Simple, requires ordinary level of skill or ability.	**S**	Unsatisfactory	**U**
				Not used or not applicable	**—**
				Combination indicates general range, i.e. Average to Good.	**A/G**

114

WOOD

Remarks	Cost	Difficulty	Appearance: interior	Appearance: exterior	Maintenance	Durability interior	Durability exterior	Fire resistivity	Thermal capacity	Insulation value	Resists wind infiltration	Resists water	Sheds water	Sheathing/decking	Beams, joists & rafters	Walls & columns	
	LM	D	P	PA	P	A	A	P	P	A	P	PA	P	P	G	E	logs
	M	LD	E	G	G	G	G	G	P	G	A	A	—	E	E	E	timber framing
•	LM	LD	—	—	—	—	—	P	P	P	A	A	—	E	E	E	ordinary framing
	L	S	PA	G	G	A	G	P	P	A	P	A	P	—	—	—	rough boards
	M	LD	E	G	A	A	G	P	P	A	A	A	P	—	—	—	dressed boards
	M	S	—	AG	A	—	G	P	P	A	E	E	A	E	—	—	plywood (exterior)
	M	S	AG	—	G	G	—	P	P	A	E	P	U	E	—	—	plywood (interior)
	M	S	PA	G	G	P	E	P	P	A	E	G	A	E	—	—	plywood, textured
	MH	S	GE	—	E	E	—	P	P	A	E	—	—	—	—	—	plywood, veneer/panelling
	LM	S	—	E	G	—	E	P	P	A	E	E	A	—	—	—	siding—clapboards
	LM	S	—	E	E	—	E	P	P	A	E	E	A	—	—	—	siding—vertical boards
	LM	LD	E	G	A	G	G	P	—	—	—	—	—	—	—	—	trim
	MH	LD	E	—	E	E	—	P	P	A	E	—	—	—	—	—	panelling
	MH	D	E	—	P	G	—	P	P	A	—	—	—	—	—	—	floors
	M	S	E	E	G	A	E	P	—	—	—	—	—	—	—	—	decking
	L	S	A	P	A	A	U	UP	P	A	E	—	—	—	—	—	particle board/fibre board
	L	S	—	—	—	—	U	UA	P	G	G	A	—	A	—	—	composition sheathing board
	M	S	—	E	E	—	E	P	P	A	E	E	E	—	—	—	shingles, siding
	MH	S	—	E	E	—	E	P	P	A	E	E	E	—	—	—	shingles, roof

• Requires finish inside & out.

Appendix II

READING LIST

1 INTRODUCTION

McHarg, Ian, **Design With Nature,** 1971. Natural History Press, Doubleday & Co., Garden City, N.Y. 11530

2 SITE SELECTION & EVALUATION

Young, Jean & Jim, **People's Guide to Country Real Estate,** 1973. Praeger, New York, N.Y. 10003

Roberts, Rex, **Your Engineered House,** 1964. M. Evans & Co., New York, N.Y. 10017

Olgyay, Victor V., **Design With Climate,** 1963. Princeton Univ. Press, Princeton, N.J. 08540

3 SITE PLANNING

Kern, Ken, **The Owner-Built Home,** 1961. Ken Kern Drafting, Oakhurst, Calif. 93644

4 PLANNING THE BUILDING

Molen, Ronald L., **House Plus Environment,** 1973. Olympus Publishing Co., Salt Lake City, Utah 84102

Watkins, A. M., **How to Judge a House,** 1972. Hawthorne Books Inc., New York, N.Y. 10016

5 PUTTING THE HOUSE TOGETHER

HUD, **In the Bank or Up the Chimney,** 1974. Department of Housing & Urban Development, Washington, D.C. 20410

Stephens, George & Alfred A. Knopp, **Remodeling Old Houses Without Destroying Their Character,** 1972. Knopf, New York, N.Y. 10022

Anderson, L. O. & Harold F. Zornig, **Build Your Own Low-Cost Home,** 1974. Peter Smith, Gloucester, Mass. 01930

Kern, Ken, **The Owner-Built Home,** 1961. Ken Kern Drafting, Oakhurst, Calif. 93644

Anderson, L. O., **How to Build a Wood Frame House,** 1973. Dover Publications, New York, N.Y. 10014

Huff, Darrell, **How to Work With Concrete & Masonry,** 1973. Barnes & Noble, New York, N.Y. 10022

BUILDING SYSTEMS 6

Edison Electric Institute, **Residential Wiring Design Guide,** 1969.
Edison Electric Institute, New York, N.Y. 10017

USDA, **Water Supply Sources for the Farmstead & Rural Home,** 1975.
(Farmers Bulletin No. 2237), USDA, Washington, D.C. 20250

Havens, David, **The Woodburner's Handbook,** 1973. Media House,
Portland, Maine 04101

USDA, **Home Heating — Systems, Fuels, Controls,** 1975. (Farmers
Bulletin No. 2235), USDA, Washington, D.C. 02050

Large, David B., **Hidden Waste: Potentials for Energy Conservation,**
1973. The Conservation Foundation, Washington, D.C. 20036

WORKING UP THE SITE 7

USDA, **Landscape for Living: Yearbook of Agriculture,** 1972. USDA,
Washington, D.C. 20250

Kramer, Jack, **Garden Planning for the Small Property,** 1972.
Chas. Scribner's Sons, New York, N.Y. 10017

Kramer, Jack, **Gardening and the Home Landscape,** 1971, Harper
& Row, New York, N.Y. 10016

Smith, Alice U. **Patios, Terraces, Decks and Roof Gardens,** 1969.
Hawthorne Books, New York, N.Y. 10016

NEW DIRECTIONS 8

Fry, L. John & Richard Merrill, **Methane Digesters for Fuel Gas and
Fertilizer,** 1973. New Alchemy Institute, Woods Hole, Mass. 02543

GARDEN WAY BOOKS

The author also recommends the following Garden Way books:

The Complete Homesteading Book, by David Robinson, 272 pp., quality paperback, $5.95; hardback, $9.95. How to live a simpler, more self-sufficient life.

Buying Country Property, by Herb Moral. 128 pp., quality paperback, $3.95. Sure to be your "best friend" when considering country property.

How to Locate in the Country: Your Personal Guide, by John Gourlie. 112 pages, quality paperback, $2.95; hardback, $4.95. Evaluates different locales and geographic areas as to suitability for you personally.

Low-Cost Pole Building Construction, by Douglas Merrilees and Evelyn Loveday. 115 pp., deluxe paperback, $4.95. This will save you money, labor, time and materials.

Build Your Own Stone House, by Karl and Sue Schwenke. 156 pp., quality paperback, $4.95; hardback, $8.95. With their help, you can build your own beautiful stone home.

New Low-Cost Sources of Energy for the Home, by Peter Clegg. 250 pp., quality paperback, $6.95; hardback, $9.95. Covers solar heating and cooling, wind and water power, wood heat and methane digestion. Packed with information.

Wood Stove Know-how, by Peter Coleman. 24 pp., illustrated paperback, $1.50. Installation, cleaning and maintenance instructions, plus much more.

The Complete Book of Heating with Wood, by Larry Gay. 128 pp., quality paperback, $3.95. Fight rising home heating costs and still keep very warm.

These Garden Way books are available at your bookstore, or may be ordered directly from Garden Way Publishing, Dept. EE, Charlotte, Vermont 05445. If your order is less than $10, please add 60¢ postage and handling.